SpringerBriefs in Electrical and Computer Engineering

Computational Electromagnetics

Series editor

Rakesh Mohan Jha, Bangalore, India

More information about this series at http://www.springer.com/series/13885

Shiv Narayan · B. Sangeetha
Rakesh Mohan Jha

Frequency Selective Surfaces based High Performance Microstrip Antenna

 Springer

Shiv Narayan
Centre for Electromagnetics
CSIR-National Aerospace Laboratories
Bangalore, Karnataka
India

Rakesh Mohan Jha
Centre for Electromagnetics
CSIR-National Aerospace Laboratories
Bangalore, Karnataka
India

B. Sangeetha
Centre for Electromagnetics
CSIR-National Aerospace Laboratories
Bangalore, Karnataka
India

ISSN 2191-8112   ISSN 2191-8120   (electronic)
SpringerBriefs in Electrical and Computer Engineering
ISSN 2365-6239   ISSN 2365-6247   (electronic)
SpringerBriefs in Computational Electromagnetics
ISBN 978-981-287-774-1   ISBN 978-981-287-775-8   (eBook)
DOI 10.1007/978-981-287-775-8

Library of Congress Control Number: 2015947811

Springer Singapore Heidelberg New York Dordrecht London

Printed on acid-free paper

Springer Science+Business Media Singapore Pte Ltd. is part of Springer Science+Business Media
(www.springer.com)

Dedicated to Dr. Sudhakar Rao

In Memory of Dr. Rakesh Mohan Jha
Great scientist, mentor, and excellent
human being

Dr. Rakesh Mohan Jha was a brilliant contributor to science, a wonderful human being, and a great mentor and friend to all of us associated with this book. With a heavy heart we mourn his sudden and untimely demise and dedicate this book to his memory.

Preface

Frequency selective surface (FSS) technology has been widely used for the design of high-performance radomes, antennas, radar absorbing structure, reflectors, etc., during the past four decades. In such applications, the FSS technology is mainly employed to enhance the performance of the candidate device/structure, and to reduce their radar signature.

High-performance low RCS (radar cross section) printed antennas are mostly preferred in stealth technology. Such printed antennas may be realized by incorporating FSS structures, either in its ground plane or as superstrate. In view of this, the design and analysis of microstrip patch antennas loaded with FSS-based (i) high impedance surface (HIS) ground plane and (ii) superstrate are presented in this book.

This brief is organized as follows: Section 1 deals with the introduction of FSS structure and Sect. 2 describes the characteristics of FSS structures. The design and analysis of microstrip antenna loaded with FSS-based HIS is discussed in Sect. 3; in this section, various types of band-stop FSS structures such as Jerusalem cross and single-square loop are designed to perform as perfect magnetic conductor (PMC), which is then used as ground plane of microstrip patch antenna (MPA). Further, the design and analysis of MPA loaded with the superstrate design, using double square loop-FSS, is studied for directivity enhancement in Sect. 4. Finally, Sect. 5 lists the conclusions of the work carried out in the book.

Shiv Narayan
B. Sangeetha
Rakesh Mohan Jha

Acknowledgments

We would like to thank Mr. Shyam Chetty, Director, CSIR-National Aerospace Laboratories, Bangalore for his permission and support to write this SpringerBrief.

We would also like to acknowledge valuable suggestions from our colleagues at the Centre for Electromagnetics, Dr. R.U. Nair, Dr. Hema Singh, Dr. Balamati Choudhury, and Mr. K.S. Venu during the course of writing this book. We express our sincere thanks to Ms. Nimisha S., Ms. Divya K.M., and Ms. Sai Samhitha, the project staff at the Centre for Electromagnetics, for their consistent support during the preparation of this book.

But for the concerted support and encouragement from Springer, especially the efforts of Suvira Srivastav, Associate Director, and Swati Meherishi, Senior Editor, Applied Sciences and Engineering, it would not have been possible to bring out this book within such a short span of time. We very much appreciate the continued support by Ms. Kamiya Khatter and Ms. Aparajita Singh of Springer toward bringing out this brief.

Contents

Frequency Selective Surfaces-Based High Performance
Microstrip Antenna   1
1   Introduction   1
2   Characteristics of FSS Structures   3
3   Microstrip Antenna Over FSS-Based High Impedance
    Ground Plane   4
    3.1   Theoretical Considerations.   5
    3.2   EM Design of Microstrip Patch Antenna Over FSS-HIS   11
    3.3   EM Performance Analysis.   12
4   Microstrip Antenna Loaded with FSS-Based Superstrate   23
    4.1   Theoretical Considerations.   24
    4.2   Estimation of Far-Field Radiation Pattern of Antenna   28
    4.3   EM Design of Microstrip Patch Antenna Loaded
          with FSS Superstrate   30
    4.4   EM Performance Analysis.   33
5   Summary   38
References   38

About the Book   41

Author Index   43

Subject Index   45

About the Authors

Dr. Shiv Narayan is with the Centre for Electromagnetics of CSIR-National Aerospace Laboratories (CSIR-NAL), Bangalore, India as Scientist, since 2008. He received his Ph.D. in Electronics Engineering from Indian Institute of Technology, Banaras Hindu University (IIT-BHU), Varanasi, India in 2006. He held the position of Scientist B between 2007 and 2008, at SAMEER (Society for Applied Microwave Electronics Engineering and Research), Kolkata, India. His research interests are broadly in the field of electromagnetics applications; these include frequency selective surfaces (FSS), metamaterials, numerical methods in electromagnetics, EM material characterization, and antennas. Dr. Shiv is the author/co-author of over 40 technical documents including peer reviewed journal and conference papers.

Ms. B. Sangeetha obtained her B.E. (ECE) degree from Visvesvaraya Technological University, Karnataka, India. She is currently a Project Engineer with the Centre for Electromagnetics of CSIR-National Aerospace Laboratories (CSIR-NAL), Bangalore, India where she works in the areas of FSS, metamaterials, and microstrip antennas.

Dr. Rakesh Mohan Jha was Chief Scientist & Head, Centre for Electromagnetics, CSIR-National Aerospace Laboratories, Bangalore. Dr. Jha obtained a dual degree in BE (Hons.) EEE and M.Sc. (Hons.) Physics from BITS, Pilani (Raj.) India, in 1982. He obtained his Ph.D. (Engg.) degree from Department of Aerospace Engineering of Indian Institute of Science, Bangalore in 1989, in the area of computational electromagnetics for aerospace applications. Dr. Jha was a SERC (UK) Visiting Post-Doctoral Research Fellow at University of Oxford, Department of Engineering Science in 1991. He worked as an Alexander von Humboldt Fellow at the Institute for High-Frequency Techniques and Electronics of the University of Karlsruhe, Germany (1992–1993, 1997). He was awarded the Sir C.V. Raman Award for Aerospace Engineering for the Year 1999. Dr. Jha was elected Fellow of

INAE in 2010, for his contributions to the EM Applications to Aerospace Engineering. He was also the Fellow of IETE and Distinguished Fellow of ICCES. Dr. Jha has authored or co-authored several books, and more than five hundred scientific research papers and technical reports. He passed away during the production of this book of a cardiac arrest.

Abbreviations

AMC	Artificial magnetic conductor
DSL-FSS	Double square loop-frequency selective surface
EBG	Electromagnetic band gap
ECM	Equivalent circuit model
EM	Electromagnetics
EMI	Electromagnetic interference
FSS	Frequency selective surfaces
HIS	High impedance surfaces
IWO	Invasive weed optimization
JC-FSS	Jerusalem crossed FSS
JC-HIS	Jerusalem cross-based HIS
MPA	Microstrip patch antenna
PEC	Perfect electric conductor
PMC	Perfect magnetic conductor
RCS	Radar cross section
TE	Transverse electric
TM	Transverse magnetic

Symbols

ε_0	Permittivity of free space
ε_r	Relative permittivity of antenna substrate
η	Intrinsic impedance
λ	Wavelength
μ_0	Permeability of free-space
μ_r	Relative permeability
Γ_{TE}, Γ_{TM}	Reflection coefficient for TE and TM mode of incidence
ω_r	Angular resonance frequency
$B(g, t)$	Capacitive susceptance
B_{TE}, B_{TM}	Capacitive susceptance for TE and TM mode of incidence
C_a	Capacitance of patch antenna
C_{1s}	Capacitance of outer square loop
C_{2s}	Capacitance of inner square loop
d	Thickness of HIS substrate
d_s	Thickness of FSS superstrate
f_r	Resonant frequency
g	Separation between adjacent JC-crosses
h	Height of microstrip antenna
k_0	Wave number
L_1	Length of microstrip antenna
L_a	Inductance of antenna
L_{1s}	Inductance of outer square loop
L_{2s}	Inductance of inner square loop
l_g	Length of JC-FSS inductive grid
L_g	Grid inductance of JC-FSS
$N(\theta)$	Refractive index
Q	Quality factor of patch antenna
R	Resistance of the patch antenna
R_D	Resistance of the FSS
w	Width of inductive grid of JC-FSS
w_c	Width of capacitive grid of JC-FSS

W_1	Width of patch antenna
W_g	Length of capacitive grid JC-FSS
X	Grid reactance of JC-FSS
X_{TE}, X_{TM}	Inductive reactance for TE and TM mode of incidence
Z_0	Characteristic impedance of free-space
Z_a	Input impedance of antenna
Z_d	Input impedance of grounded dielectric slab
Z_s	Input impedance of FSS-HIS
Z_{FSS}	Input impedance of FSS

List of Figures

Figure 1   Typical FSS types and their frequency response characteristics; **a** array of metallic patches shows low-pass behavior, **b** array of apertures on conducting screen shows high-pass behavior, **c** array of metallic loops shows band-stop behavior, **d** array of aperture loops on conducting screen shows band-pass behavior 4

Figure 2   **a** Schematic of rectangular microstrip patch antenna over FSS-based HIS, **b** unit cell of Jerusalem cross FSS. 6

Figure 3   **a** Equivalent circuit of the rectangular microstrip patch antenna, **b** equivalent circuit of the Jerusalem cross FSS, **c** equivalent circuit of the FSS-based HIS 7

Figure 4   Input impedance of rectangular microstrip patch antenna. *Red lines* show computed results at CEM, CSIR-NAL. *Blue lines* show reported results (Volakis 2007) 12

Figure 5   **a** Input impedance of rectangular microstrip patch antenna designed at 10 GHz. **b** Return loss of rectangular MPA versus frequency (designed at 10 GHz). 13

Figure 6   Reflection phase of the JC-FSS-based HIS. *Solid blue line* shows computed result at CEM based on equivalent circuit model. *Dotted red line* shows reported result based on numerical model (Hosseinipanah and Wu 2009). 14

Figure 7   Real and imaginary parts of input impedance of JC-FSS-based HIS 14

Figure 8   Return loss of rectangular MPA microstrip antenna with PEC ground plane and FSS-HIS ground plane designed at 5.8 GHz. Bullet points show reported results (Monavar and Komjani 2011). *Solid lines* show computed results at CEM 15

Figure 9   Return loss of the proposed MPA with PEC and FSS-based HIS ground plane designed at 10 GHz 16

Figure 10   Return loss of rectangular MPA with FSS-HIS ground plane
            for different length of the inductive grid (l_g) of JC-FSS
            element.   17
Figure 11   Return loss of rectangular MPA with FSS-HIS ground plane
            for different gap between the adjacent crosses (g) of FSS
            element.   17
Figure 12 E-plane radiation pattern of rectangular microstrip antenna
            with PEC ground plane and JC-FSS-based HIS ground
            plane   18
Figure 13   **a** Schematic of square loop FSS-based HIS, **b** unit cell of
            SSL-FSS.   19
Figure 14   Equivalent circuit of single square loop FSS-based HIS   19
Figure 15   Real and imaginary parts of input impedance of single
            square loop FSS-based HIS.   20
Figure 16   Return loss of rectangular MPA with PEC ground plane and
            SSL-FSS-based HIS ground plane   21
Figure 17 E-plane radiation pattern of rectangular MPA with PEC
            ground plane and SSL-FSS-based HIS ground plane.   21
Figure 18   Real and imaginary parts of input impedance of
            SSL-FSS-based HIS (Teflon substrate)   22
Figure 19   Return loss of rectangular MPA with PEC ground plane and
            SSL-FSS-based HIS (Teflon substrate) ground plane   22
Figure 20 E-plane radiation pattern of rectangular MPA with PEC
            ground plane and SSL-FSS-based HIS (Teflon substrate)
            ground plane.   23
Figure 21   **a** Schematic of rectangular microstrip patch antenna loaded
            with FSS superstrate, **b** geometry of unit cell of
            DSL-FSS   25
Figure 22   **a** Equivalent circuit of rectangular microstrip patch antenna,
            b equivalent circuit of DSL-FSS   26
Figure 23   Equivalent circuit of the antenna loaded with square loop
            FSS-based superstrate.   27
Figure 24   Equivalent transmission line of MPA covered with FSS
            superstrate.   28
Figure 25   Transmission characteristics of DSL-FSS structure;
            a reported (Lou etΓÇÖal. 2005), **b** computed at CEM   31
Figure 26   Transmission characteristics of proposed DSL-FSS structure
            at different incidence angles (0°, 30°, and 45°) for TE
            polarizations   32
Figure 27   Reflection characteristics of proposed DSL-FSS structure at
            different incidence angles (0°, 30°, and 45°) for TE
            polarizations   32
Figure 28   EM characteristics of MPA loaded with superstrate; **a** input
            impedance, and **b** return loss.   34

Figure 29   Validation of **a** E-plane, and **b** H-plane pattern of rectan-
            gular MPA. *Dotted black lines* show reported results
            (Volakis 2007). *Solid blue lines* show computed results at
            CEM   35
Figure 30   Radiation pattern of rectangular MPA and MPA covered
            with DSL-FSS superstrate; **a** E-plane and **b** H-plane.   36
Figure 31   Radiation pattern of rectangular microstrip antenna and
            MPA covered with FSS-based superstrate by keeping air gap
            between them; **a** E-plane and **b** H-plane   37

Frequency Selective Surfaces-Based High Performance Microstrip Antenna

Abstract In order to fulfill the growing demand high-performance low RCS antenna in stealth technology, FSS-based antenna is found to be the better candidate. In view of this, the design and analysis of microstrip patch antennas (MPA) loaded with (i) FSS-based HIS (high-impedance surface) ground plane and (ii) FSS-based superstrate are presented in this book with proper formulations and graphical presentations.

1 Introduction

Frequency selective surfaces (FSS) technology have potential applications in various sectors such as aerospace, medical, microwave industry, and real estate. In medical sector, FSS is used to prevent electromagnetic interference (EMI) in sophisticated medical instruments such as MRI, implantable medical devices, etc. (Gupta and Gururaj 2013). In microwave industry, it is used to reduce the leakage of power through oven door by incorporating the reflection-type FSS structure into the walls of the oven cavity. In real estate sector, FSS is employed to facilitate better wireless communications and to reduce EMI/EMC due to unwanted signals (Philippakis et al. 2004). In aerospace sector, FSS is mainly employed to enhance the performance and reduce the radar cross section (RCS) of various devices such as radome, RAS, reflector antennas, antennas, microwave circuits, etc.

Recently, high performance low observable antennas have demanded applications in strategic area. This can be accomplished by incorporating the FSS structure to the planar antenna, where FSS is employed either as *superstrate* or *high impedance ground plane* of antenna. The incorporation of FSS to the antenna mainly enhances its gain, bandwidth, and reduces its out-of-band RCS (Pirhadi et al. 2012; Li et al. 2010). The planar antennas such as dipoles, microstrip patches, etc., need a ground plane, which acts as a reflector to enhance the radiation gain. However, the metallic ground plane is one of the most important scattering components of the antenna, because it largely reflects the energy of incident waves

© The Author(s) 2016
S. Narayan et al., *Frequency Selective Surfaces based High Performance Microstrip Antenna*, SpringerBriefs in Computational Electromagnetics,
DOI 10.1007/978-981-287-775-8_1

outside the operating bandwidth. In order to reduce the scattering component outside the operating bandwidth, the conventional ground plane can be replaced with a stop-band FSS (Lu et al. 2009) as it works as perfect reflector within the band and completely pass the signal outside the band.

In view of this, Lu et al. (2009) designed a low RCS antenna by replacing its ground plane with novel stop-band FSS structure which works as high impedance ground plane surface. Li et al. (2010) presented a low RCS and high gain reflect-array antenna backed on FSS structure. The proposed antenna is reported to exhibit "in-band" improvement of 1.1 dB, reduction in side-lobe level by 6.4 dB, and significant reduction in out-of-band RCS. Further, a low profile low RCS array antenna was designed by replacing its PEC ground plane with hybrid FSS structure (Genovesi et al. 2012) without altering in-band radiation characteristics. The proposed antenna structure was analyzed based on *periodic method of moment*. Recently, a microstrip antenna backed by a novel hybrid ground plane consisting of miniaturized FSS elements with partial metallic plane is presented for out-of-band RCS reduction over wide frequency range 1–20 GHz (Yang et al. 2013) at oblique angle of incidence. In order to enhance the bandwidth of the antenna, Yeo et al. (2002) designed, a novel FSS-based electromagnetic bandgap (EBG) structure using Genetic algorithm and used it as a ground plane of planar antenna to enhance the bandwidth of the antenna. Further, the bandwidth of a microstrip antenna has been enhanced by using Jerusalem crossed FSS (JC-FSS)-based artificial magnetic conductor (AMC) ground plane (Monavar and Komjani 2011), where invasive weed optimization (IWO) algorithm were implemented to optimize the FSS structure and antenna for optimal performance. The proposed JC-FSS-based antenna was analyzed using full-wave analysis method.

Another application of FSS is the design of superstrate for the antenna to enhance its radiation characteristics. In this view, Lee et al. (2004) analyzed a microstrip antenna covered with FSS-based superstrate to enhance the directivity of the antenna. A significant enhancement in directivity was achieved from 17.29 to 24.92 dBi for three different unit cell dimensions of FSS as compared to the conventional patch antenna. Pirhadi et al. (2007) demonstrated various types of FSS-based superstrate designed by single layer and multilayered square loop structure to enhance the directivity of dual-band EBG resonator antenna. Later, gain enhancement of various types of antennas were presented in open domain using FSS-based superstrate (Foroozesh and Shafai 2010; Chiu and Chen 2011; Pirhadi et al. 2012).

In this brief, the design and analysis of microstrip patch antenna (MPA) is presented loaded with (i) various types of FSS-based high impedance ground plane and (ii) FSS-based superstrate. The analysis of composite antenna structure is carried out using transmission line equivalent circuit model as it requires less processing time and memory as compared to the methods based on full-wave analysis. The details of this work have been discussed in the following sections.

2 Characteristics of FSS Structures

The FSS is a periodical structure, which has specific reflection and transmission characteristics for the electromagnetic waves passing through it. The FSS structures resonate at a designed frequency and attain spectral selectivity (Loui 2006). Basically, FSS structures can be divided into two categories based on element geometry. The first type comprises of metallic patches on the substrate, which is usually referred to as *capacitive* FSS. Such type of FSS structures exhibit low-pass filter characteristics. The other type of FSS structure comprises of apertures on a metallic screen, which is commonly known as an *inductive* FSS. This type of FSS configurations shows high-pass filter characteristics. Generally, FSS has two major applications in aerospace; one application of FSS is to design antenna radomes to enhance transmission efficiency within the band and sharp roll-off characteristics outside the operating band. Another is to employ FSS in the design of high performance reflector antenna systems. Apart from these applications, FSS structure is recently used to design artificial magnetic conductors such as HIS, EBG ground plane, to enhance the radiation characteristics and reduce the structural RCS of the antenna.

The transmission type FSS can further be divided into thick or thin, depending on the thickness of metallic screen (Chen and Stanton 1991). If the thickness of the FSS (metallic sheet) is less than 0.001λ, the FSS is considered as a "thin"filter, which is modeled as *infinitely thin* in the numerical modeling. If the thickness of metallic sheet is greater than 0.001λ, the FSS is modeled as a "thick" filter. A thick-metal FSS finds applications, where mechanical strength and power handling are important factors. Further, the FSS characteristics can be divided into four categories namely; *low-pass*, *high-pass*, *band-pass*, and *band-stop*, based on their frequency responses as shown in Fig. 1.

From Fig. 1a, b, it is evident that the low-pass and high-pass FSS structures are complementary surfaces. It means that they cover the entire surface jointly (Gustafsson et al. 2005). Moreover, the transmission and reflection properties of low-pass and high-pass FSS structures are complimentary to each other as per Babinate's principle. This concept is also applicable to the band-pass and band-stop FSS structure as shown in Fig. 1c, d. However, such type of complementary relationship is applicable to only "thin FSS" structures that do not have dielectric backing layer. With the dielectric backing, the resonance frequency of FSS structure is shifted to lower side by an amount $1/\sqrt{\varepsilon_r}$, where ε_r is the relative permittivity of dielectric.

Fig. 1 Typical FSS types and their frequency response characteristics; **a** array of metallic patches shows low-pass behavior, **b** array of apertures on conducting screen shows high-pass behavior, **c** array of metallic loops shows band-stop behavior, **d** array of aperture loops on conducting screen shows band-pass behavior

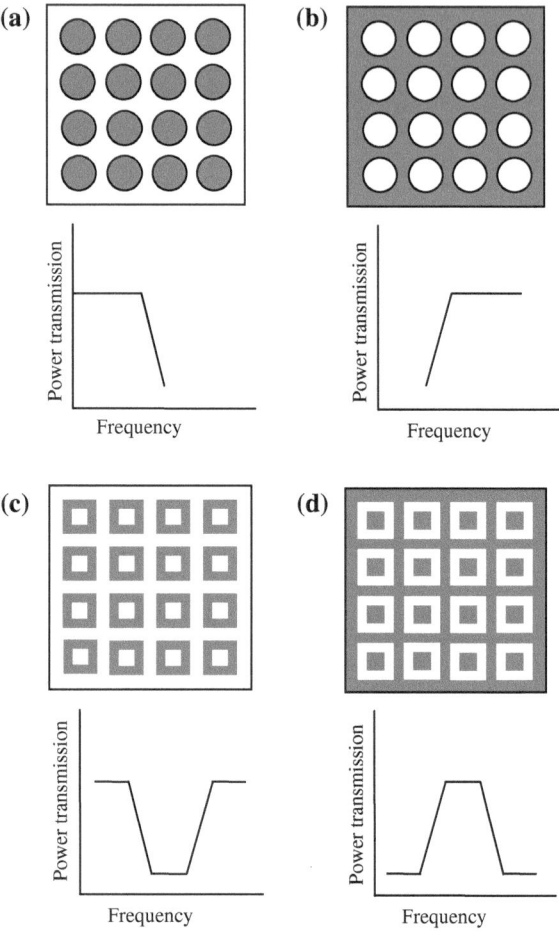

3 Microstrip Antenna Over FSS-Based High Impedance Ground Plane

The band-pass FSS is severally used as radome for antennas to enhance its performance and reduce the RCS of the antenna (Narayan et al. 2012; Costa; and Monarchio 2012). In contrast, stop-band FSS has been used as high impedance ground plane in planar antennas to enhance its gain, bandwidth, and out-of-band RCS reduction (Lu et al. 2009; Genovesi et al. 2012). Antennas such as dipoles, microstrip patches, etc., need a ground plane, which works as a reflector to enhance the radiation gain. But, the metallic ground plane is one of the most important scattering components of the antenna because it largely reflects the energy of incident waves. In order to reduce the scattering component and hence to enhance the radiation gain of the microstrip antenna, the conventional ground plane can be replaced with a stop-band FSS.

Further, it was reported that the bandwidth of conventional *MPA* can be enhanced by removing its PEC ground plane with perfect magnetic conductor (PMC) (Monavar and Komjani 2011). In this case, the image of electric current is in-phase and parallel to the original current distribution in contrast to PEC ground plane. So the antenna impedance matching would be possible over a relatively wide frequency range. Artificial magnetic conductors (AMC) exhibit the behavior of a PMC at resonance, which are also called as high impedance surfaces (HIS) or EBG ground planes. Such types of structures are used to enhance the radiation characteristics of antenna and to reduce the effect of surface waves (Hosseini and Hakkak 2008). Generally, the artificial magnetic conductors are designed with FSS backed by a grounded dielectric (Hosseinipath and Wu 2009).

In this section, the analysis of a rectangular *MPA* is presented over HIS substrate designed using different types of FSS structure such as on Jerusalem crossed FSS, single square loop FSS, etc. Basically, FSS-based HIS acts as PMC ground plane for the antenna. The EM analysis of proposed microstrip antenna is carried out based on equivalent circuit model as it requires less memory and CPU time as compared to full-wave analysis method (Monavar and Komjani 2011). The proposed antenna exhibits a significant enhancement of impedance bandwidth (19.8 %) as compared with the conventional patch antenna (10.2 %) at 10 GHz.

3.1 Theoretical Considerations

In this work, a rectangular MPA is considered for the theoretical simulation. The side view of a rectangular patch antenna loaded with HIS ground plane is shown in Fig. 2, where Jerusalem crossed FSS (stop-band) backed by grounded dielectric acts as high impedance ground plane for the antenna.

A general microstrip antenna has a ground plane on one side of a dielectric substrate and a metallic radiating patch on the other side of it. The patch can be fed through a coaxial line or microstrip line to excite the antenna. According to modal-expansion cavity model, MPA is considered as a thin TM_z-mode cavity having magnetic walls around the peripheral of the patch and electric walls at the top and bottom of the patch (Carver and Mink 1981). As antenna is excited, the fringing field is formed between the ground plane and periphery of the patch that leads to the radiation from the patch antenna. This is due to the fact that the dimensions of the patch are finite along its length and width. As a result of fringing phenomenon, the electrically length of the patch increases and hence its physical dimensions increases. Let us consider the extension of the length on each side represented by ΔL. A practical approximation for the normalized extension of length ($\Delta L/h$) is given by Balanis (1997)

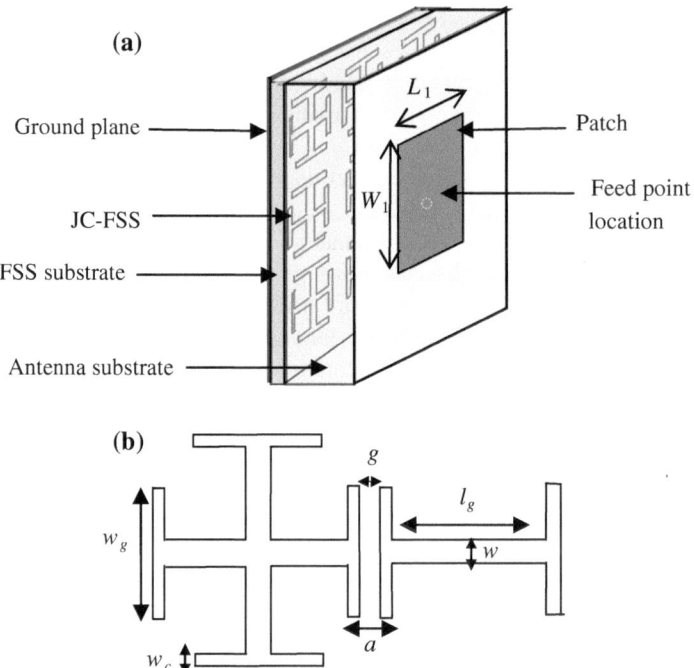

Fig. 2 **a** Schematic of rectangular microstrip patch antenna over FSS-based HIS, **b** unit cell of Jerusalem cross FSS

$$\frac{\Delta L}{h} = 0.412 \frac{(\varepsilon_e + 0.3)\left(\frac{W_1}{h} + 0.264\right)}{(\varepsilon_e - 0.258)\left(\frac{W_1}{h} + 0.8\right)} \tag{1}$$

where, h and W_1 are the height of the substrate and width of the antenna, respectively. ε_e is the effective dielectric constant of the antenna substrate, expressed as

$$\varepsilon_e = \frac{\varepsilon_r + 1}{2} + \frac{\varepsilon_r - 1}{2}\left[1 + 12\frac{h}{W_1}\right]^{-\frac{1}{2}} \tag{2}$$

The width W_1 of the patch antenna is determined by

$$W_1 = \frac{c}{2f_r}\sqrt{\frac{2}{(\varepsilon_e + 1)}} \tag{3}$$

where, c represents the velocity of light in free-space and f_r denotes the resonant frequency of the microstrip antenna. Since the length of the patch is extended by ΔL on each side, the effective length of the patch is

$$L_{\text{eff}} = L_1 + 2\Delta L \tag{4}$$

where, L_1 represents the actual length of the rectangular MPA.

The analysis of proposed antenna is carried out based on cavity model in combination with equivalent circuit model. Accordingly, a rectangular MPA can be represented by a parallel *RLC* resonant circuit as shown in Fig. 3a, where, R_a represents the resistance due to the ohmic losses in the metallic parts of the patch. L_a and C_a represent the inductance and capacitance due to the magnetic and electric energy stored in the antenna, respectively. The proposed microstrip antenna is excited with coaxial probe (50 Ω) feed, which offers a series inductance L_p to the *RLC* circuit of patch antenna (Fig. 3a). The expression for L_p is given by (Kanaujia and Viswakarma 2006)

$$L_p = \frac{\eta_0 h}{2\pi c} \ln\left[\frac{4c}{\zeta \omega d \sqrt{\varepsilon_r}}\right] \tag{5}$$

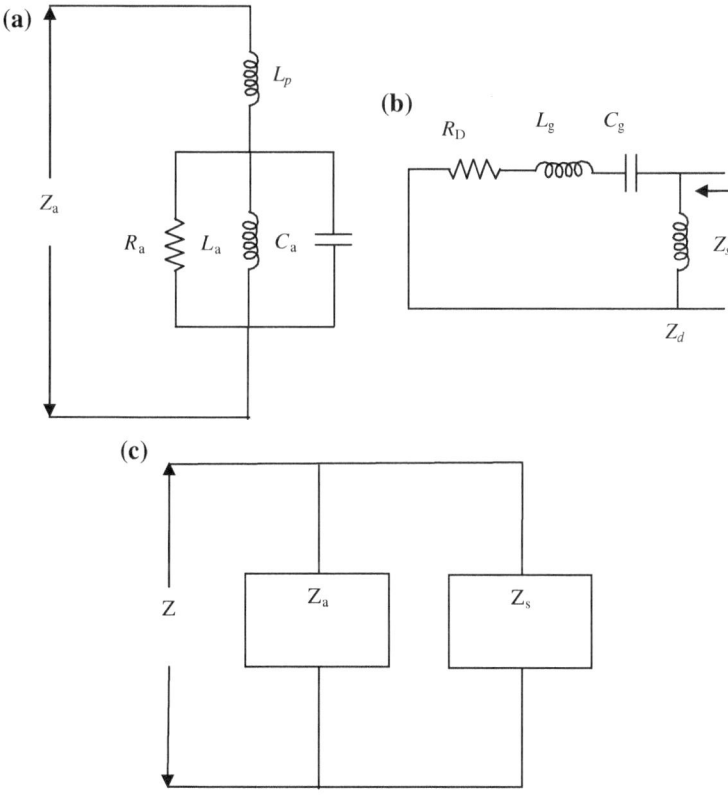

Fig. 3 **a** Equivalent circuit of the rectangular microstrip patch antenna, **b** equivalent circuit of the Jerusalem cross FSS, **c** equivalent circuit of the FSS-based HIS

where, $\zeta = 1.781072\ldots$, $i = 1$ is determined from Euler's constant. d is the diameter of coaxial probe and c represents the velocity of light in free-space. η_0 is the free-space impedance.

Now from Fig. 3a, the input impedance of rectangular MPA can be determined as

$$Z_a = \frac{1}{\frac{1}{R_a} + j\omega C_a + \frac{1}{j\omega L_a}} + j\omega L_p \tag{6}$$

where, R_a is the equivalent resistance due to ohmic losses in the metallic parts of the patch. L_a and C_a are the equivalent inductance and capacitance, respectively, corresponding to magnetic and electric energy stored within the patch antenna. The expressions for R_a, L_a, and C_a is given as (Bahl and Bhartia 1980)

$$C_a = \frac{\varepsilon_e \varepsilon_0 L_1 W_1}{2h} \cos^{-2}\left(\frac{\pi y_0}{L_1}\right) \tag{7}$$

where, y_0 represents the length of the feed-point along the length of the patch antenna,

$$L_a = \frac{1}{C_a \omega_r^2}, \tag{8}$$

and

$$R_a = \frac{Q}{\omega_r C_a} \tag{9}$$

where, Q is the total quality factor of the microstrip antenna (Derneryed and Lind 1979) and ω_r is the angular resonance frequency of MPA.

In this work, the *Jerusalem crossed FSS* (JC-FSS) is used to design the high impedance ground plane for the proposed antenna. The unit cell of the Jerusalem cross FSS is comprised of capacitive and inductive elements (Fig. 2b) and it exhibits band-stop characteristics. For this structure, the surface impedance plays a vital role in determining the resonant frequency and the phase of reflection coefficient (Hosseinipanah and Wu 2009). The equivalent circuit of Jerusalem crossed FSS backed by grounded dielectric is shown in Fig. 3b, where R_D, L_g, and C_g represent the resistance, inductance, and capacitance, respectively, associated with dielectric backed Jerusalem cross FSS. Z_d is the impedance offered by the grounded dielectric.

The inductive reactance of the Jerusalem cross FSS is given by Hosseini and Hakkak (2008) as

$$X_g = Z_0 \tan\left(\frac{kl_g}{2}\right) \tag{10}$$

where, l_g is the length of the inductive grid and Z_0 is the characteristic impedance of Jerusalem cross strip. k is the wave number, expressed as

$$k = \omega\sqrt{\mu_0\varepsilon_0}\sqrt{\varepsilon_{\text{reff}}} \tag{11}$$

where, $\varepsilon_{\text{reff}}$ is the effective relative permittivity of HIS substrate and is determined as

$$\varepsilon_{\text{reff}} = \frac{\varepsilon_r + 1}{2} + \frac{\varepsilon_r - 1}{2}\left(1 + 10\frac{d}{W_g}\right)^{-\frac{1}{2}} \tag{12}$$

where, W_g is the length of the capacitive grid and d is the height of the HIS substrate.

The capacitance offered by Jerusalem cross array can be determined, based on the capacitance between two parallel patches placed apart on a dielectric slab as

$$C_g = \frac{2W_g}{\pi}\varepsilon_0\varepsilon_{\text{reff}}\cosh^{-1}\left(\frac{a}{g}\right) \tag{13}$$

Since, the electric field lines associated with a lossy medium surrounding the FSS structure lead to the dielectric loss. Such loss component can be expressed as a series resistor in parallel with lossless capacitor (between the adjacent elements) and is given by Costa and Monarchio (2012)

$$R_D = \frac{-2\varepsilon_r''}{\omega C_0\left(\varepsilon_r' + 1\right)^2} \tag{14}$$

where, ε_r' and ε_r'' are the real and imaginary parts of the complex relative permittivity, respectively. C_0 represents the capacitance of the FSS structure in free-standing configuration, is given as

$$C_0 = \frac{2W_g}{\pi}\varepsilon_0\cosh^{-1}\left(\frac{a}{g}\right) \tag{15}$$

The lumped impedance offered by JC-FSS printed on lossy substrate can be expressed as

$$Z_{\text{FSS}} = R_s + jX_s \tag{16}$$

where, R_s and X_s is the resistance and reactance, respectively offered by FSS structure.

The input impedance of a PEC-backed dielectric slab of thickness d is given as (Costa and Monarchio 2012)

$$Z_d = j \frac{\eta_0}{\sqrt{\varepsilon_r' + j\varepsilon_r''}} \tan\left(k_0 \sqrt{\varepsilon_r' + j\varepsilon_r''} d\right) \tag{17}$$

where, η_0 is the free-space impedance and k_0 is the free-space wave number. From Eq. (17), the real and imaginary parts of the input impedance of the grounded dielectric can be expressed as

$$\operatorname{Re}\{Z_d\} \cong \frac{\zeta_0}{\sqrt{\varepsilon_r'}} \left[\frac{\varepsilon_r''}{2\varepsilon_r'} \tan\left(k_0 d \sqrt{\varepsilon_r'}\right) - \left(k_0 d \frac{\varepsilon_r''}{2\sqrt{\varepsilon_r'}}\right) \times \left(1 + \tan^2\left(k_0 d \sqrt{\varepsilon_r'}\right)\right) \right] \tag{18}$$

and

$$\operatorname{Im}\{Z_d\} \cong \frac{\zeta_0}{\sqrt{\varepsilon_r'}} \left[\tan\left(k_0 d \sqrt{\varepsilon_r'}\right) \right] \tag{19}$$

The input impedance of the FSS-HIS structure is determined by the parallel combination of impedance of JC-FSS and surface impedance of grounded dielectric slab (Fig. 3b) is expressed as

$$Z_s = Z_d || Z_{\text{FSS}} = \frac{Z_d Z_{\text{FSS}}}{Z_d + Z_{\text{FSS}}} \tag{20}$$

Now, the input impedance of proposed microstrip antenna designed over HIS substrate can be estimated by equivalent circuit modelas shown in Fig. 3c. It is apparent that the input impedance of proposed microstrip antenna can be determined by the parallel combination of HIS impedance and the MPA impedance. In order to calculate the input impedance of MPA, the height of the HIS substrate is also added to the height of the conventional MPA. Thus, the input impedance of the HIS-based antenna can be expressed as

$$Z = Z_a || Z_s = \frac{Z_a Z_s}{Z_a + Z_s} \tag{21}$$

Using Eq. (21), the return loss of the proposed antenna can be computed as

$$R = 20 \log|\Gamma| \tag{22}$$

where, Γ is the reflection coefficient of the proposed antenna given as

$$\Gamma = \frac{Z - Z_c}{Z + Z_c} \tag{23}$$

where, Z_c is the characteristic impedance of coaxial feed line (50 Ω).

3.2   EM Design of Microstrip Patch Antenna Over FSS-HIS

Since the proposed antenna comprises of a rectangular MPA whose ground plane is replaced by FSS-based HIS. The EM design considerations of conventional rectangular patch antenna and FSS-based HIS is discussed separately in the following subsections.

3.2.1   EM Design of Microstrip Patch Antenna

In this work, a rectangular MPA is designed at the operating frequency of 10 GHz. The design parameters of the antenna are determined based on cavity model. The height of the antenna substrate and its dielectric constant is considered to be 1.588 mm and 2.2, respectively. The length and width of the patch are calculated to be 9.06 and 11.86 mm, respectively.

The rectangular MPA is excited with a coaxial probe (50 Ω) located at $x_0 = 0.312$ mm (along the length of MPA) and $y_0 = $ W/2. For efficacy of cavity model and equivalent circuit approach, the input impedance of a rectangular MPA is compared with that of reported result at 1.575 GHz (Volakis 2007). The reported rectangular MPA was designed at resonant frequency 1.575 GHz and its designed parameters are; length $L = 62.55$ mm, width $W = 93.83$ mm, dielectric constant of the substrate $\varepsilon_r = 2.2$, substrate height $h = 1.524$ mm. The reported antenna was excited with a coaxial probe of radius 0.635 mm and located at $x_0 = 18.5$ mm and $y_0 = $ W/2. A good agreement is achieved between computed and reported results as shown in Fig. 4. Further, the input impedance and return loss of proposed rectangular MPA is computed and their frequency response is also studied as shown in Fig. 5a, b. It is observed that the proposed rectangular MPA antenna resonates at 10 GHz with perfect feed matching.

3.2.2   EM Design of FSS-Based HIS

The HIS structure consists of Jerusalem cross FSS backed by grounded dielectric substrate in this work. The Jerusalem cross FSS is intended to design at the same frequency as that of rectangular MPA (10 GHz), for band-stop characteristics.

The designed parameters of the JC-FSS-based HIS are; width of the inductive component, $w = 0.1$ mm, length of the inductive component, $l_g = 4.0$ mm, length of the capacitive component, $W_g = 3.43$ mm, width of the capacitive segment, $w_c = 0.29$ mm, and separation between the adjacent crosses, $g = 0.38$ mm. The height of the HIS substrate is considered to be, $d = 0.34$ mm.

The impedance of FSS-based HIS is also determined based on equivalent circuit model. In order to validate the approach, the reflection phase of the Jerusalem crossed FSS-based HIS is computed based on equivalent circuit model as discussed in the previous section for the designed parameters of $W_g = 3.5$ mm, $w_c = 0.1$ mm, $g = 0.4$ mm, $l_g = 4$ mm, $w = 0.1$ mm, $d = 1$ mm, and $\varepsilon_r = 2.2$. The reflection phase of

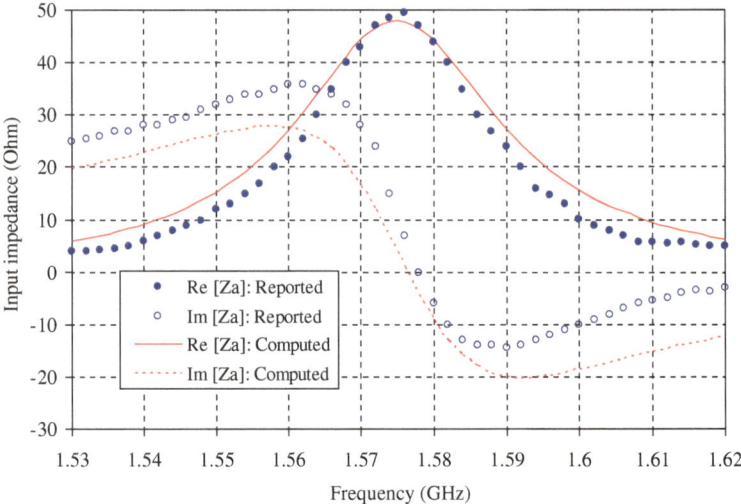

Fig. 4 Input impedance of rectangular microstrip patch antenna. *Red lines* show computed results at CEM, CSIR-NAL. *Blue lines* show reported results (Volakis 2007)

the JC-FSS-based HIS is studied with respect to operating frequency and compared with reported result estimated based on numerical simulation as shown in Fig. 6. It is observed that excellent agreement is obtained between computed and reported result (Hosseinipanah and Wu 2009).

Further, the impedance of the proposed FSS-based HIS is determined using equivalent circuit model and studied its frequency response as shown in Fig. 7. It is observed that the proposed FSS-based HIS resonates at 10 GHz and exhibits very high impedance at resonance. Thus, the Jerusalem crossed FSS backed by grounded dielectric behaves as HIS at resonance and hence it can be used as ground plane for microstrip antenna.

3.3 EM Performance Analysis

The EM analysis of microstrip antenna loaded with various FSS-based HIS such as Jerusalem cross FSS and square loop FSS are carried out based on equivalent transmission line approach. The details are discussed in the following subsections.

3.3.1 Jerusalem Cross FSS-Based Microstrip Antenna

This subsection deals with the EM performance analysis of rectangular MPA over JC-FSS-based HIS followed by validation of equivalent circuit approach.

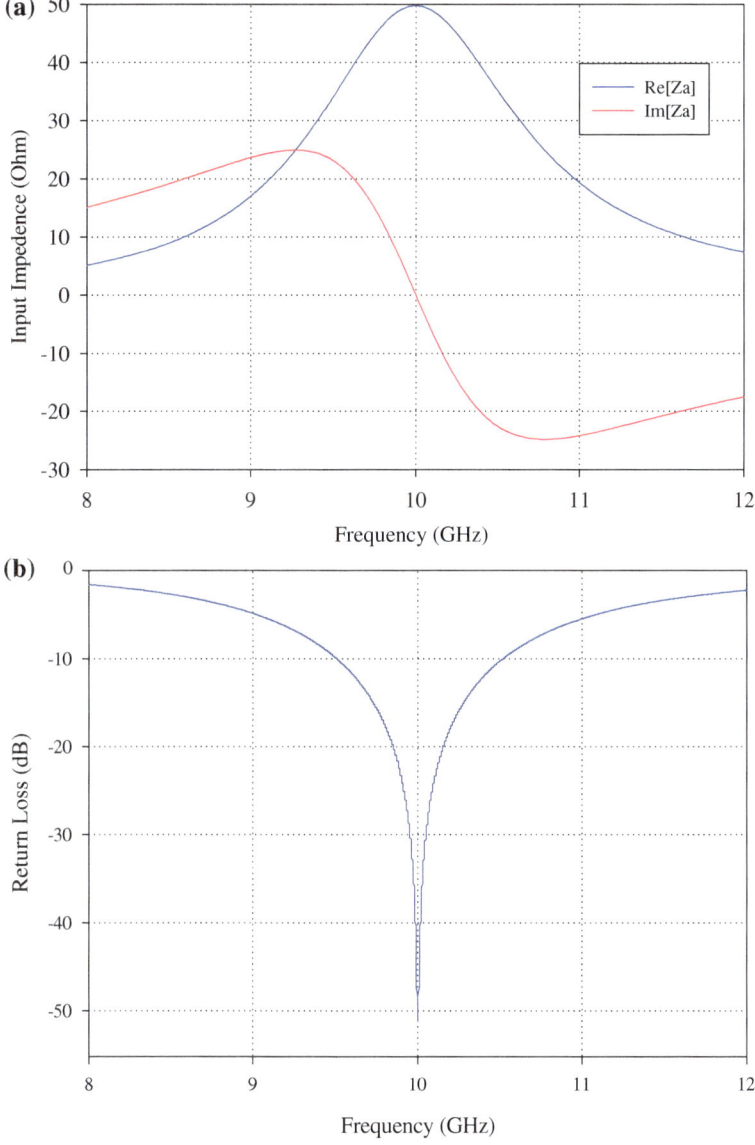

Fig. 5 **a** Input impedance of rectangular microstrip patch antenna designed at 10 GHz. **b** Return loss of rectangular MPA versus frequency (designed at 10 GHz)

Validation of equivalent circuit approach: The EM analysis of microstrip antenna loaded with FSS-based HIS is carried out based on equivalent circuit model. For efficacy of this novel approach, a rectangular MPA loaded with Jerusalem cross FSS-based HIS is investigated based on equivalent circuit model

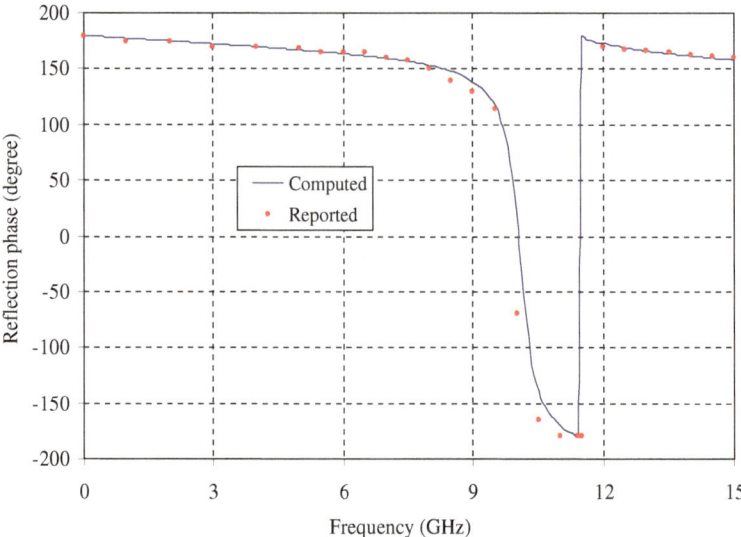

Fig. 6 Reflection phase of the JC-FSS-based HIS. *Solid blue line* shows computed result at CEM based on equivalent circuit model. *Dotted red line* shows reported result based on numerical model (Hosseinipanah and Wu 2009)

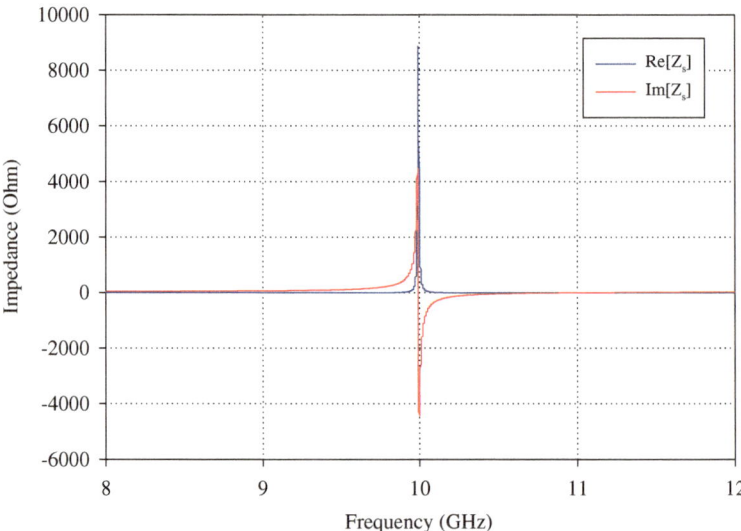

Fig. 7 Real and imaginary parts of input impedance of JC-FSS-based HIS

and its result is compared with that of reported result (Monavar and Komjani 2011) for the designed parameters; length of the microstrip antenna, $L_1 = 12.5$ mm, width of the patch, $W_1 = 17.5$ mm, height of antenna substrate, $h = 1.58$ mm, width of inductive grid, $w = 0.55$ mm, length of capacitive grid, $w_g = 8.5$ mm, length of the inductive grid, $l_g = 11.64$ mm, distance between two inductive grids, $a = 4.31$ mm, and height of FSS-HIS substrate = 3.16 mm, at designed frequency of 5.8 GHz.

The reported antenna which was simulated based on full-wave method, exhibited a bandwidth of 3.44 % with PEC ground plane and 10.41 % on FSS-based HIS ground plane as shown in Fig. 8. While the rectangular MPA which is simulated based on equivalent circuit approach, exhibits a bandwidth of 4.3 % with PEC ground plane and enhanced bandwidth of 11.89 % on FSS-based HIS ground plane (Fig. 8). It is observed that both approaches show almost similar bandwidth enhancement of ~ 7 % as compared to the conventional rectangular MPA. However, in equivalent circuit approach, the resonance frequency of HIS-based antenna shifted to 5.71 GHz instead of 5.8 GHz, which may be due to the approximation of lumped parameters considered in this approach.

Performance analysis of JC-FSS-based antenna: Finally, the EM performance analysis of the proposed antenna carried out in this work is based on equivalent circuit approach. For the analysis, both rectangular MPA and FSS-based HIS is designed at the center frequency of 10 GHz. The return loss of proposed microstrip antenna is computed and compared with that of conventional rectangular MPA as shown in Fig. 9. It is noted that the proposed microstrip antenna exhibits an impedance bandwidth (-10 dB) from 8.65 to 10.63 GHz, i.e., 19.8 % with HIS substrate, while it shows 10.2 % for PEC ground plane. It is obvious that the impedance bandwidth of the proposed antenna with HIS ground plane is enhanced

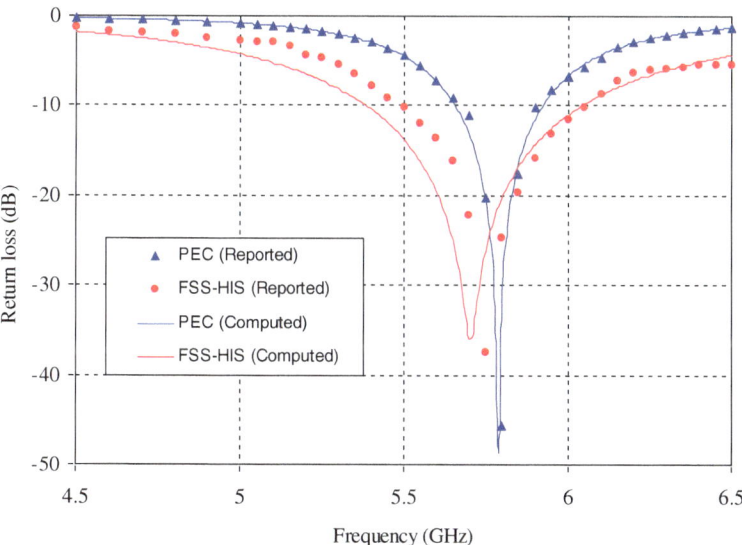

Fig. 8 Return loss of rectangular MPA microstrip antenna with PEC ground plane and FSS-HIS ground plane designed at 5.8 GHz. Bullet points show reported results (Monavar and Komjani 2011). *Solid lines* show computed results at CEM

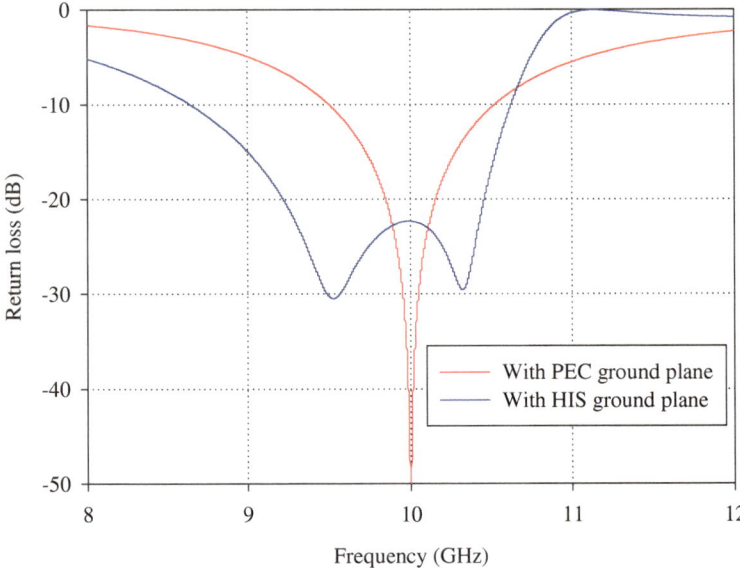

Fig. 9 Return loss of the proposed MPA with PEC and FSS-based HIS ground plane designed at 10 GHz

by 9.6 % as compared to PEC ground plane. Thus, the Jerusalem crossed FSS-based HIS enhances the bandwidth of a rectangular MPA.

Further, the effect of geometrical parameters of FSS elements on return loss of antenna is studied by varying the length of the inductive grid (l_g) and gap between the adjacent crosses (g) of JC-FSS. It is observed that as the length of the inductive grid increases, the bandwidth and resonance frequency of the proposed antenna decreases as shown in Fig. 10. While, the bandwidth and resonance frequency of proposed antenna increases with the increase in the gap between the adjacent crosses of JC-FSS (Fig. 11). Thus, the bandwidth of the proposed HIS-based antenna can be tuned by varying the geometrical parameters of JC-FSS elements such as l_g and g.

The far-field pattern of proposed antenna is estimated using the principles of reciprocity theorem and equivalent transmission line analogy (Volakis 2007). The details of this approach are discussed in Sect. 4. Accordingly, the proposed antenna problem reduces to the scattering of plane wave on the grounded multilayered structure and its reflection coefficient is determined based on equivalent transmission line method as discussed in Sect. 3.1.

The far-field radiation pattern of proposed antenna is determined in E-plane and compared with the rectangular MPA having PEC ground plane (Fig. 12). It is observed that the FSS-HIS-based antenna shows a significant enhancement of beamwidth (13.18°) as compared to that of conventional microstrip antenna.

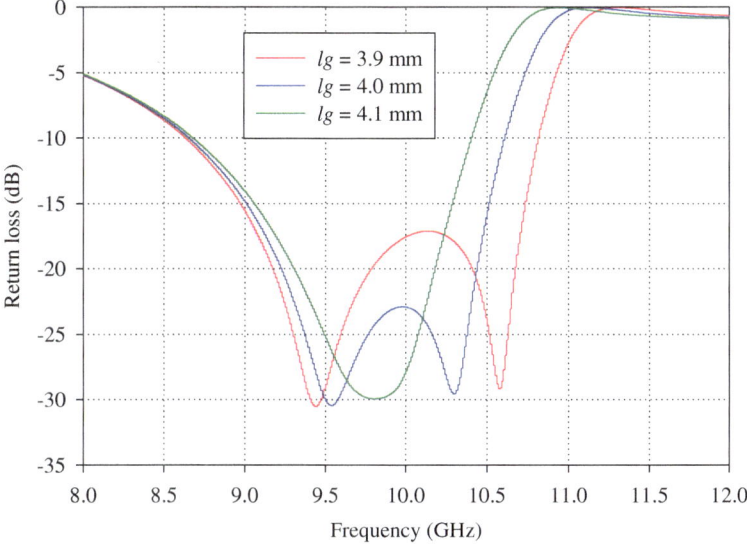

Fig. 10 Return loss of rectangular MPA with FSS-HIS ground plane for different length of the inductive grid (l_g) of JC-FSS element

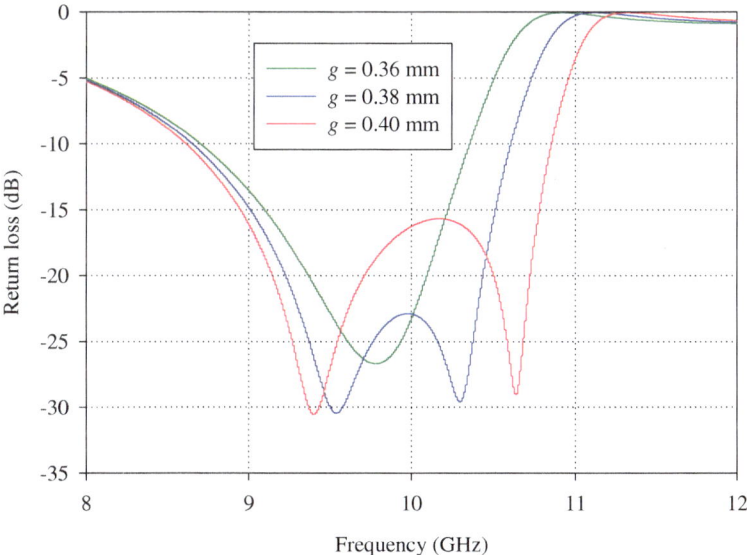

Fig. 11 Return loss of rectangular MPA with FSS-HIS ground plane for different gap between the adjacent crosses (g) of FSS element

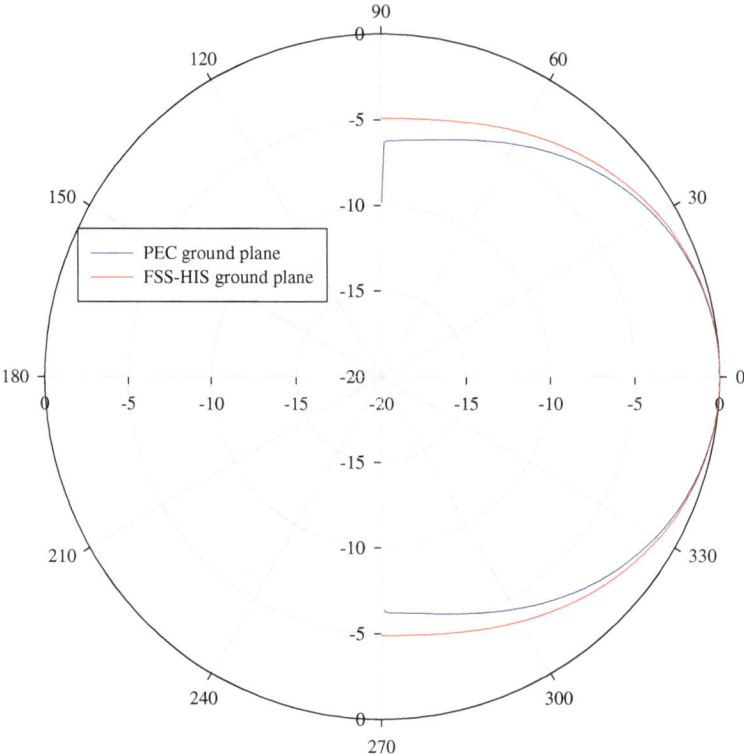

Fig. 12 *E*-plane radiation pattern of rectangular microstrip antenna with PEC ground plane and JC-FSS-based HIS ground plane

3.3.2 Square Loop FSS-Based Microstrip Antenna

The performance analysis of proposed microstrip antenna is further studied by using single square loop FSS (SSL-FSS)-based HIS ground plane. The schematic of square loop FSS-based HIS is shown in Fig. 13, where p, d, and s represent the periodicity, side length of square loop, and thickness of square loop grid, respectively. The equivalent circuit of square loop FSS backed by grounded dielectric which forms the HIS is shown in Fig. 14. Here, R_D, L_s, and C_s are associated with the dielectric loss around the square loop, magnetic current, and gap between the square loops, respectively. Z_d represents the impedance offered by grounded dielectric beneath the SL-FSS. The numerical value of R_D, L_s, C_s, and Z_d are determined using the same concept as considered for JC-FSS HIS. Similar to the JC-FSS case, the single square loop FSS-based HIS is also designed at 10 GHz.

The designed dimensions of square loop FSS-based HIS are given in Table 1. The input impedance of SSL-FSS-based HIS are computed based on equivalent circuit model. From Fig. 15, it is evident that the square loop FSS backed by

Fig. 13 a Schematic of
square loop FSS-based HIS,
b unit cell of SSL-FSS

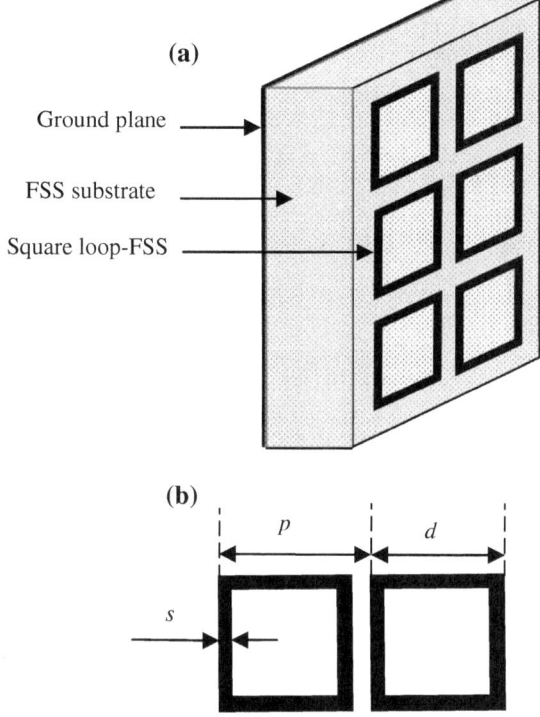

(a)

Ground plane

FSS substrate

Square loop-FSS

(b)

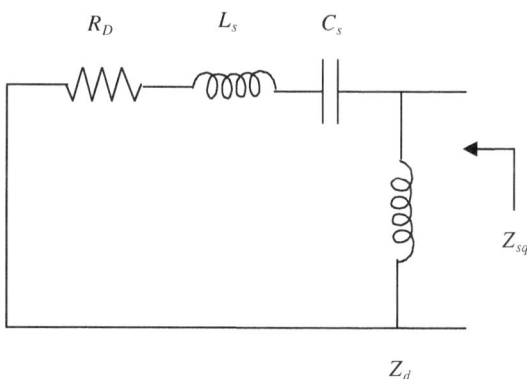

Fig. 14 Equivalent circuit of
single square loop FSS-based
HIS

grounded dielectric ($\varepsilon_r = 2.2$ and tan $\delta = 0.0009$) reveals very high impedance
(7134 Ω) at resonance. Thus, it can be used as HIS ground plane for microstrip
antenna.

Further, the MPA is loaded over SSL-FSS-based HIS and is analyzed based on
transmission line equivalent circuit model. The return loss of SSL-FSS-based
antenna is studied with respect to operating frequency and compared with that of

Table 1 Designed
parameters of single square
loop FSS-based HIS

Designed parameters	Optimized value
Periodicity of square loop (p)	7.22 mm
Thickness of square loop grid (s)	0.22 mm
Length of square grid (d)	6.0 mm
Distance between two inductive grids (a)	1.66 mm
Height of HIS substrate	0.36 mm
Dielectric constant of HIS substrate	2.2

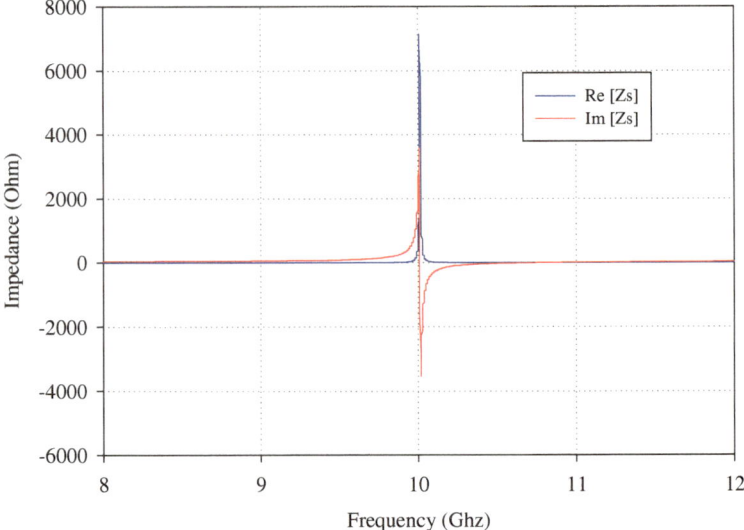

Fig. 15 Real and imaginary parts of input impedance of single square loop FSS-based HIS

conventional rectangular MPA as shown in Fig. 16. It is observed that the micro-strip antenna designed over SL-FSS-based HIS ground plane shows a bandwidth (10 dB) enhancement of 6.06 % as compared with that of conventional MPA.

The radiation characteristic of proposed antenna is determined based on reciprocity theorem and transmission line theory. The computed radiation characteristic is compared with that of conventional rectangular MPA. It is found that over the single square loop FSS-based HIS ground plane, MPA exhibits a significant enhancement of beamwidth (11.4°) in E-plane as compared to that of conventional MPA. This is evident from Fig. 17. However, the resonance frequency of FSS-HIS loaded microstrip antenna slightly shifted to 9.75 GHz apart from 10 GHz, which may be due to the approximate considerations of lumped parameters in equivalent circuit model.

In order to further study the effect of dielectric constant of HIS substrate on the EM performance of proposed microstrip antenna, the dielectric material of proposed SSL-FSS-based HIS is replaced with Teflon ($\varepsilon_r = 2.08$ and tan $\delta = 0.001$) and then composite antenna structure is analyzed based on equivalent circuit model.

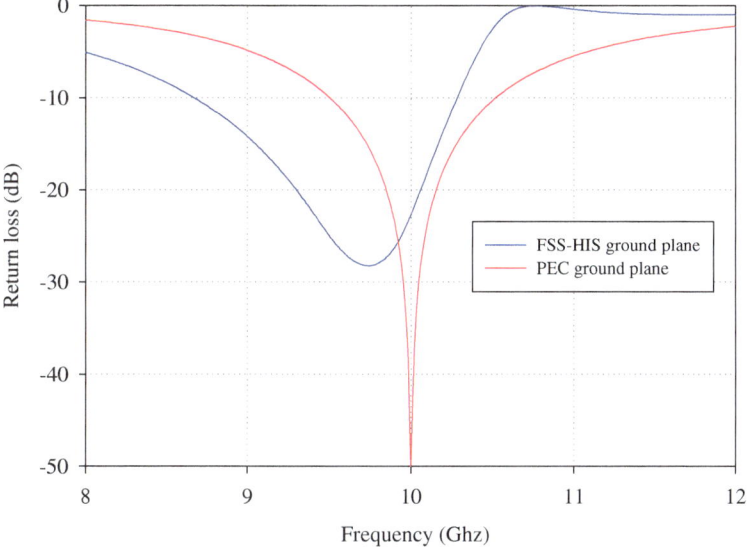

Fig. 16 Return loss of rectangular MPA with PEC ground plane and SSL-FSS-based HIS ground plane

Fig. 17 *E*-plane radiation pattern of rectangular MPA with PEC ground plane and SSL-FSS-based HIS ground plane

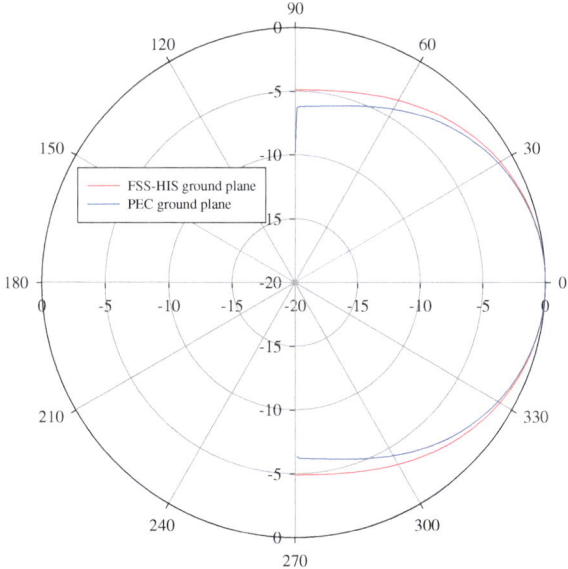

The input impedance of SSL-FSS-based HIS is studied *with respect to* operating frequency. It is evident from Fig. 18 that the Teflon-based HIS exhibits very high impedance (7364 Ω) at resonance as compared to that of RT-Duroid-based

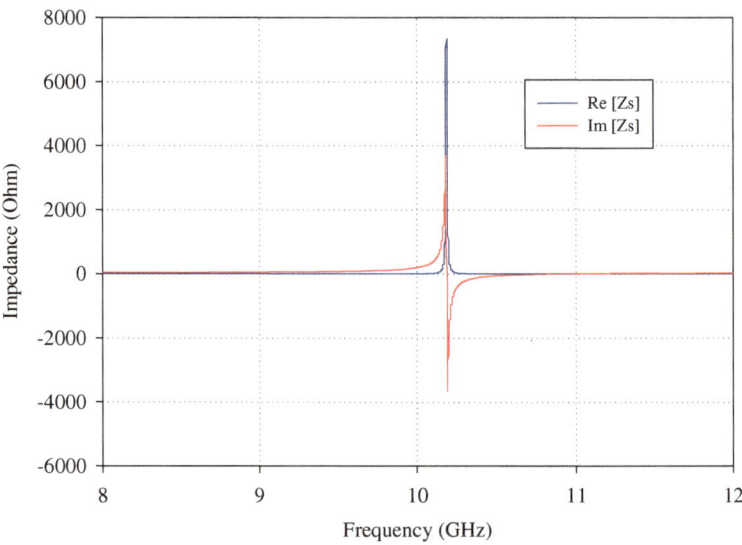

Fig. 18 Real and imaginary parts of input impedance of SSL-FSS-based HIS (Teflon substrate)

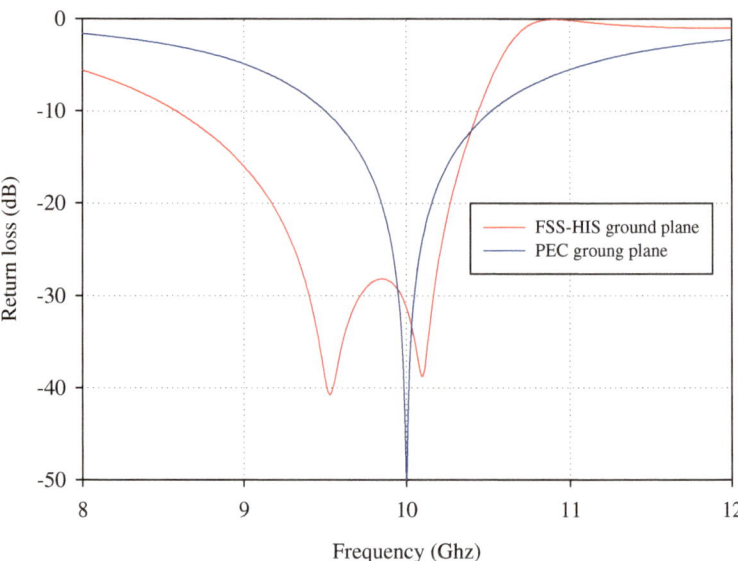

Fig. 19 Return loss of rectangular MPA with PEC ground plane and SSL-FSS-based HIS (Teflon substrate) ground plane

SSL-FSS-based HIS. Further, the return loss of composite antenna is studied *with respect to* operating frequency and compared with that of conventional MPA as shown in Fig. 19. It is observed that the MPA over Teflon substrate HIS reveals a

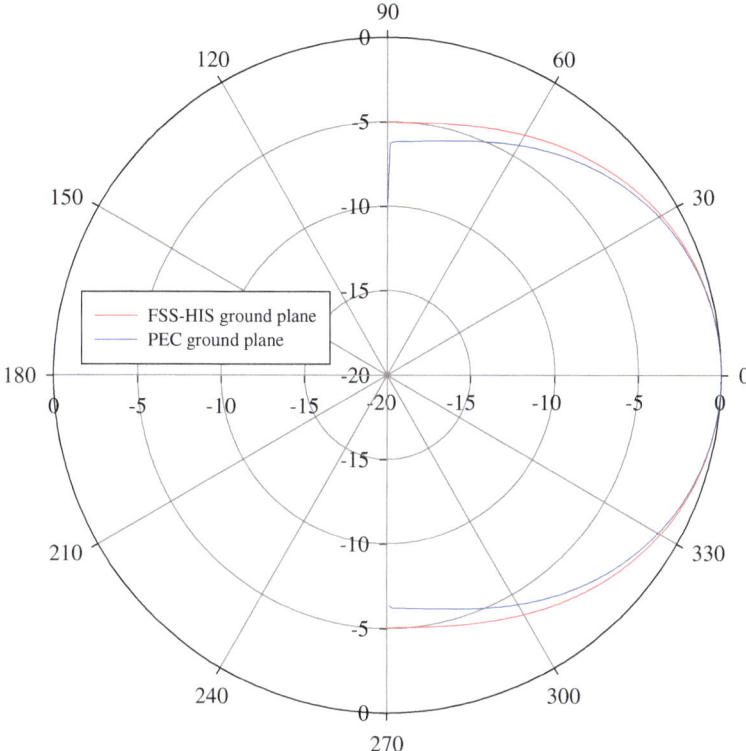

Fig. 20 *E*-plane radiation pattern of rectangular MPA with PEC ground plane and SSL-FSS-based HIS (Teflon substrate) ground plane

significant enhancement of bandwidth (8.4 %) as compared to the conventional MPA. This result concludes that the bandwidth of proposed antenna can further be enhanced by changing the dielectric properties of HIS substrate. The radiation characteristics of composite antenna loaded with Teflon-based HIS is also studied as shown in Fig. 20. This shows a significant enhancement in beamwidth (12.6°) as compared to conventional MPA.

4   Microstrip Antenna Loaded with FSS-Based Superstrate

A superstrate layer is generally used to enhance the directivity as well as bandwidth of the microstrip antenna (Jackson and Alexopoulos 1985; Pirhadi et al. 2006). In order to design low profile low RCS antenna, the superstrate must be compact and easily fabricated. For the design of such antenna, frequency selective surfaces are found to be the best suitable candidate for the design of superstrate (Foroozesh and

Shafai 2010; Pirhadi et al. 2012) as it exhibits the filter characteristics for the EM wave impinging on it. The bandwidth, polarization, and radiation characteristics (e.g., side lobe level, directivity etc.) of antenna can be controlled using FSS superstrate or its combination with reactive surfaces (Foroozesh and Shafai 2006; Rodes et al. 2007). In addition, the FSS superstrate can also be used as a polarizer (Pirhadi et al. 2012).

Several methods have been proposed in open domain for the analysis of microstrip antenna loaded with superstrate such as FEM, MoM, etc., (Alexopoulos and Jackson 1984; Pirhadi et al. 2007). However, the analyzes based on full-wave methods are computationally complex and require large CPU time and memory to converge the solution. In this endeavor, the EM analysis of a MPA covered with FSS-superstrate is presented based on transmission line *equivalent circuit model* (ECM) as it is computationally less complex and requires less CPU time and memory to handle such problems.

The radiation characteristic of proposed antenna is estimated using transmission line analogy and reciprocity theorem. Here, double square loop-frequency selective surface (DSL-FSS) has been utilized to design the FSS-based superstrate layer. Since, the DSL-FSS is basically a double resonant structure that provides a transmission band formed due to combination of two reflection bands. This means that the DSL-FSS provides both transmission and reflection bands, which are insensitive to the angle of incidence (Luo et al. 2005). The proposed antenna exhibits a directivity enhancement of 3.85 dB in *E*-plane and 4.06 dB in *H*-plane as compared to that of conventional microstrip antenna.

4.1 Theoretical Considerations

A rectangular MPA is considered for the theoretical simulation in this work. The side view of rectangular MPA covered with FSS-based superstrate is shown in Fig. 21, where the DSL-FSS backed by a dielectric substrate is used to design the superstrate. The superstrate loaded antenna is fed through a 50 Ω coaxial cable to excite the field within the antenna. According to cavity model theory, a rectangular MPA structure can be analyzed by solving a parallel *RLC* resonant circuit as shown in Fig. 22a, where R_a represents the resistance due to the ohmic losses in the metallic parts of the patch. L_a and C_a represent the inductance and capacitance due to the magnetic and electric energy stored in the antenna, respectively. The input impedance of antenna is determined using Eq. (6).

In this work, the DSL-FSS structure is used to design the superstrate for the proposed microstrip antenna. The geometry of DSL-FSS is shown in the Fig. 21b, which consists of concentric inner and outer rings separated by a gap in between. According to the equivalent circuit model, a DSL-FSS structure can be represented as parallel combination of two series *LC* resonant circuits as shown in Fig. 22b,

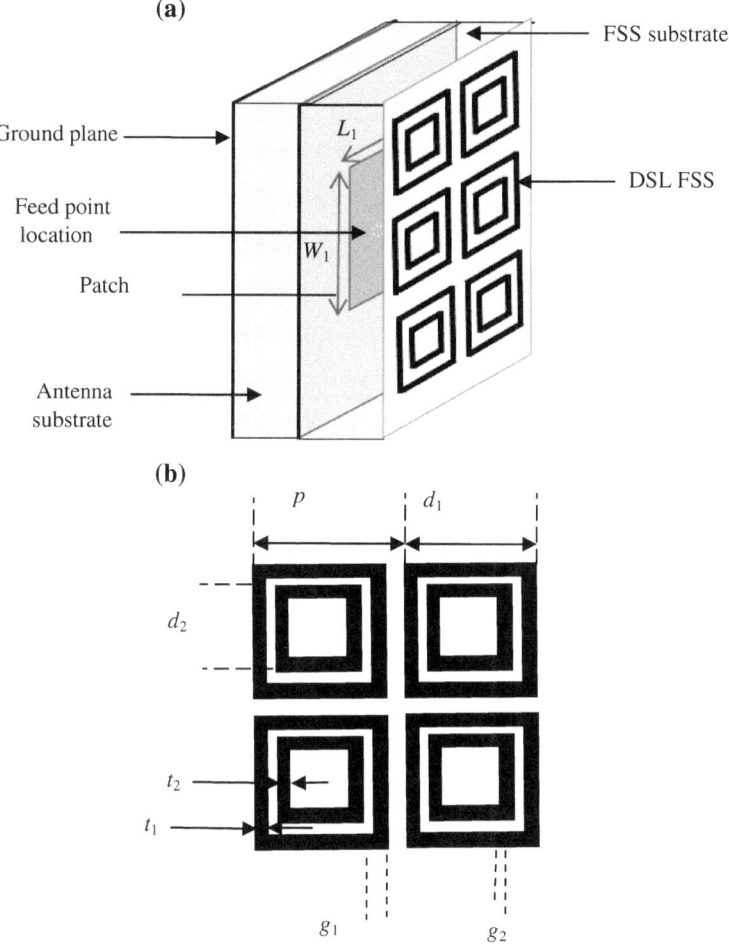

Fig. 21 a Schematic of rectangular microstrip patch antenna loaded with FSS superstrate, **b** geometry of unit cell of DSL-FSS

where L_{1s} and C_{1s} are the inductance and capacitance, respectively offered by the outer square ring. L_{2s} and C_{2s} represent the inductance and capacitance, respectively offered by the inner square ring.

The numerical value of L_{1s}, C_{1s}, L_{2s}, and C_{2s} can be determined by the expressions (Luo et al. 2005), given below

$$X_1 = \omega L_{1s} = 2.0 \times \frac{X_{1s}X_{2s}}{X_{1s} + X_{2s}} \times \left(\frac{d_1}{p}\right) \tag{24}$$

Fig. 22 a Equivalent circuit of rectangular microstrip patch antenna, **b** equivalent circuit of DSL-FSS

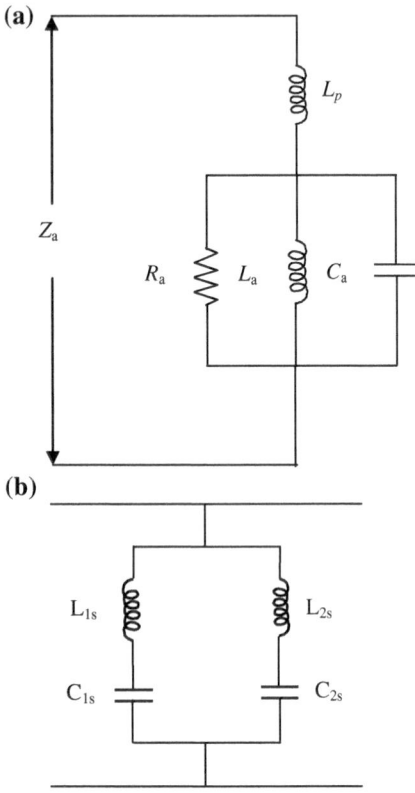

where, $X_{1s} = F(p, t_1, \lambda)$ and $X_{2s} = F(p, t_2, \lambda)$

$$X_2 = \omega L_{2s} = F(p, 2t_2, \lambda) \times \left(\frac{d_2}{p}\right) \tag{25}$$

where, $F(p, 2t_2, \lambda)$ is determined from (Lee and Langley 1985)

$$B_1 = \omega C_{1s} = 0.75 \times B_{1s} \times \left(\frac{d_1}{p}\right) \tag{26}$$

where, $B_{1s} = 4.0 \times \varepsilon_{\text{eff}} \times F(p, g_1, \lambda)$.

Here, ε_{eff} is the effective dielectric constant and it is given by Costa et al. (2012)

$$\varepsilon_{\text{eff}} = \varepsilon_{r\_h} + (\varepsilon_{r\_h} - 1)\left[\frac{-1}{\exp^N(x)}\right] \tag{27}$$

where,

$$\varepsilon_{r\_h} = \frac{\varepsilon_r + 1}{2},$$

(28)

$$x = \frac{10\,d}{p},$$

(29)

and N is an exponential factor in which numerical value is 1.3 for ring-like structure and 1.8 for the cross-like structure. d and p represent the thickness of the FSS substrate and periodicity of the FSS unit cell, respectively.

$$B_2 = \omega C_{2s} = \frac{B_{1s}B_{2s}}{B_{1s} + B_{2s}} \times \left(\frac{d_2}{p}\right)$$

(30)

where, $B_{2s} = 4.0 \times \varepsilon_{eff} \times F(p, g_2, \lambda)$.

Since in practice, the inductive and capacitive reactances offered by FSS structure are different for TE and TM mode of incidence wave. These reactances can be determined separately for TE and TM polarizations (Lee and Langley 1985).

Now the admittance of DSL-FSS is determined as

$$Y = j\left[\left(\frac{B_1}{1 - X_1B_1}\right) + \left(\frac{B_2}{1 - X_2B_2}\right)\right]$$

(31)

The impedance offered by DSL-FSS-based superstrate can be computed by

$$Z_s = \frac{1}{Y}$$

(32)

Finally, the input impedance of proposed antenna (i.e., microstrip antenna covered with FSS superstrate) shown in the Fig. 23 is determined by the parallel

Fig. 23 Equivalent circuit of the antenna loaded with square loop FSS-based superstrate

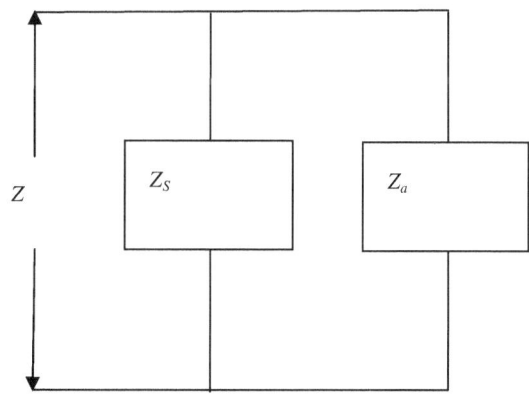

combination of the impedance of MPA and the impedance of FSS superstrate (DSL-FSS) expressed as

$$Z = Z_a || Z_s = \frac{Z_a Z_s}{Z_a + Z_s} \qquad (33)$$

Using Eq. (33), the input impedance and return loss of MPA loaded with FSS superstrate can be determined. This is to be noted that the above expression will only be used to determine the input impedance and return loss of composite antenna when there would not be any gap between the superstrate and antenna. For the air gap between the antenna superstrate, the additional impedance offered by gap will be used in parallel to the antenna impedance.

4.2   Estimation of Far-Field Radiation Pattern of Antenna

The radiation pattern of antenna is determined based on transmission line theory and reciprocity theorem. Here, the MPA loaded with FSS superstrate is considered as a transmission line. According to the transmission line theory, the impedance of the whole structure can be determined by considering the antenna and FSS as a section of the transmission line as show in Fig. 24. Each section of transmission line is analyzed by its characteristic impedance and propagation constant that basically depends on the dielectric properties of the layer and angle of incidence (Jackson and Alexopoulos 1985). Thus, the input impedance of each layer is estimated and used as the load for the preceding layer is explained below.

In Fig. 24, the first section represents the microstrip antenna, where the shorted load denotes the ground plane of the antenna and Z_{c1} is the characteristic impedance

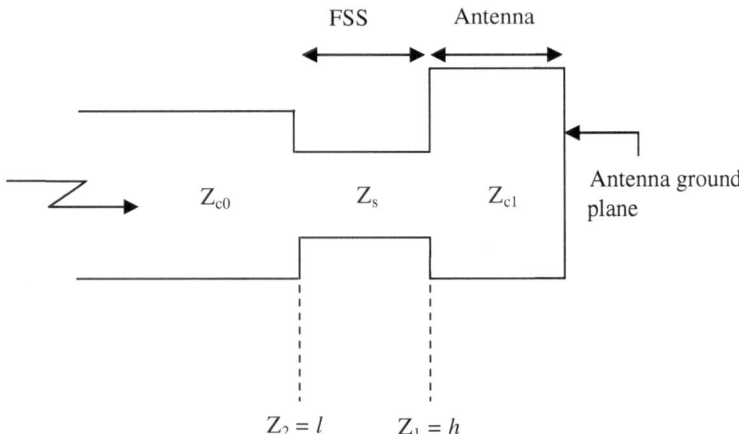

Fig. 24 Equivalent transmission line of MPA covered with FSS superstrate

of the antenna. The expressions of Z_{c1} for both TE and TM polarization is given by Jackson and Alexopoulos (1985) as

$$Z_{c1} = \frac{\eta_0 N_1(\theta)}{\varepsilon_1}, \quad \text{For TE-polarization} \tag{34}$$

$$Z_{c1} = \frac{\eta_0 \mu_1}{N_1(\theta)}, \quad \text{For TM-polarization} \tag{35}$$

The impedance at the terminal plane $Z = h$ can be determined by transmission line theory as

$$Z_1 = j Z_{c1} \tan(k_0 N_1(\theta) h) \tag{36}$$

where, k_0 and $N_1(\theta)$ are the propagation constant and refractive index corresponding to the antenna substrate, respectively. Similarly, the input impedance at the terminal plane $Z = l$ can be computed by

$$Z_2 = Z_s \frac{Z_1 + j Z_s \tan(\beta l)}{Z_s + j Z_1 \tan(\beta l)} \tag{37}$$

where, Z_s is the characteristic impedance of FSS superstrate which can be computed by Eq. (32). l represents the height of the FSS substrate. Finally, the reflection coefficient of the whole structure can be determined as

$$\Gamma = \frac{Z_2 - Z_{c0}}{Z_2 + Z_{c0}} \tag{38}$$

where, Z_{c0} represents the characteristic impedance of free-space, which can be expressed for both TE and TM polarizations (Jackson and Alexopoulos 1985) as

$$Z_{c0} = \eta_0 \cos(\theta), \quad \text{For TE-polarization} \tag{39}$$

$$Z_{c0} = \eta_0 \sec(\theta), \quad \text{For TM-polarization} \tag{40}$$

Thus, the reflection coefficient of proposed antenna is determined using Eq. (38), which will be used for the estimation of far-field radiation pattern of antenna both in E-plane and H-plane as discussed below.

The far-field pattern of proposed antenna is estimated using the principles of reciprocity theorem. Accordingly, the far-field $E_{i=0,\phi}(r, \theta, \phi)$ of antenna can be determined by placing a unit-amplitude testing dipole at the far-field distance in the direction of interest (θ or ϕ). Thus, the proposed antenna problem reduces to the scattering of plane wave on the grounded multilayered structure and its reflection coefficient is determined based on transmission line analogy as discussed above.

The far-field radiation pattern of antenna loaded with superstrate for both E- and H-plane can be computed by Volakis (2007)

$$E_{\theta}^{\text{patch}}(r, \theta, \phi) = -2Wh \left(\frac{E_0}{\eta_0}\right) \cos \phi \left(1 - \Gamma^{\text{TM}}(\theta)\right) \cos \left(k_x \frac{L}{2}\right)$$
$$\times \sin c \left(k_y \frac{L}{2}\right) \tan c(k_{Z1}h) \tag{41}$$

$$E_{\varphi}^{\text{patch}}(r, \theta, \phi) = 2Wh \left(\frac{E_0}{\eta_0}\right) (\cos \theta \sin \phi) \left(1 - \Gamma^{\text{TE}}(\theta)\right) \cos \left(k_x \frac{L}{2}\right)$$
$$\times \sin c \left(k_y \frac{W}{2}\right) \tan c(k_{Z1}h) \tag{42}$$

where,

$$E_0 = \left(\frac{-j\omega\mu_0}{4\pi R}\right) e^{-jk_0 R} \tag{43}$$

and

$$k_x = k_0 \sin \theta \cos \phi \tag{44}$$

$$k_y = k_0 \sin \theta \sin \phi \tag{45}$$

$$k_{z1} = k_0 N_1(\theta) \tag{46}$$

$$N_1(\theta) = \sqrt{\varepsilon_r \mu_r - \sin^2 \theta} \tag{47}$$

where, the term Γ^{TE} and Γ^{TM} denotes the reflection coefficients in TE and TM mode, respectively, η_0 is the free-space impedance, R is the far-field distance measured from the center of the patch. $N_1(\theta)$ is the refractive index of the medium.

4.3 EM Design of Microstrip Patch Antenna Loaded with FSS Superstrate

A rectangular MPA is designed in this work at the operating frequency of 10 GHz. The design parameters of the antenna are determined based on cavity model. The height of the antenna substrate and its dielectric constant is considered to be 1.588 mm and 2.2, respectively. The length and width of the patch are calculated to be 9.06 and 11.86 mm, respectively. The microstrip antenna is excited with a coaxial probe of 50 Ω located at the feed length, $y = 3.126$ mm. The equivalent circuit method is first employed to calculate the input impedance and then the return loss of the MPA.

Further, the FSS-based superstrate is designed at the same operating frequency (10 GHz) using transmission line equivalent circuit approach. In this work, DSL has been used to design the FSS-based superstrate, which shows the pass-band characteristic at the desired frequency for both TE and TM polarizations. The optimized design parameters of DSL-FSS structure are: periodicity of array, p = 8.102 mm, width of the outer square loop, w_1 = 0.48 mm, width of the inner loop, w_2 = 0.48 mm, gap between two outer square loop, g_1 = 0.162 mm, gap between inner and outer square loop, g_2 = 0.58 mm, length of the outer square loop, d_1 = 7.939 mm, and length of the inner square loop, d_2 = 6.779 mm.

For efficacy of the equivalent circuit approach, the transmission characteristics of proposed DSL-FSS is studied and compared with the reported measured results at normal incidence (Luo et al. 2005). A good agreement is obtained between the computed and reported result as shown in the Fig. 25. Further, the transmission characteristic of proposed DSL-FSS is studied at various angles of incidence (0°, 30°, and 45°) for TE polarizations as shown in Fig. 26. It is observed that the designed FSS reveals excellent transmission efficiency of 10 GHz for both polarizations.

This is also corroborated from reflection data as shown in Fig. 27 for TE polarizations, which shows very low (almost zero) reflection efficiency at different incidence angles corresponding to the desired frequency. Thus, the proposed DSL-FSS is found to be best suitable candidate for the design of superstrate to enhance the directivity of MPA.

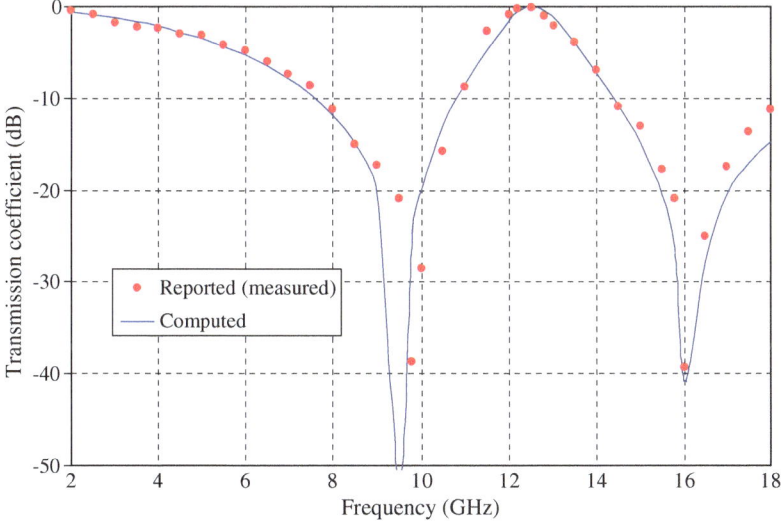

Fig. 25 Transmission characteristics of DSL-FSS structure; **a** reported (Luo et al. 2005), **b** computed at CEM

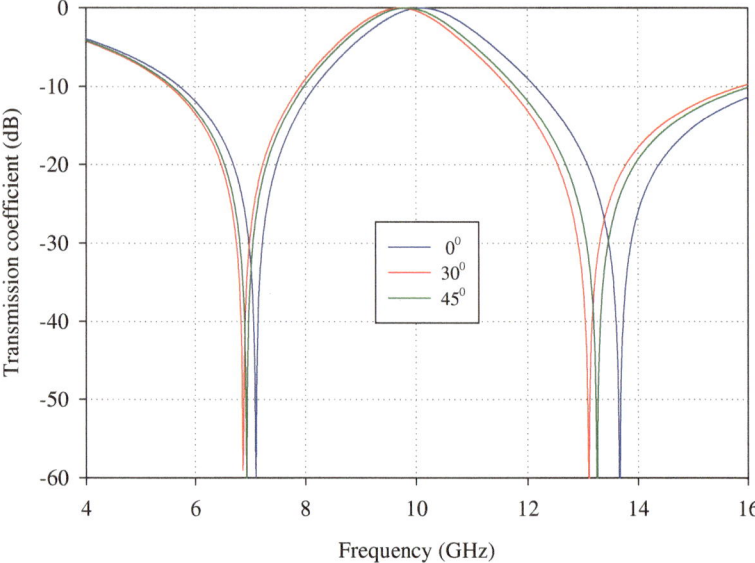

Fig. 26 Transmission characteristics of proposed DSL-FSS structure at different incidence angles (0°, 30°, and 45°) for TE polarizations

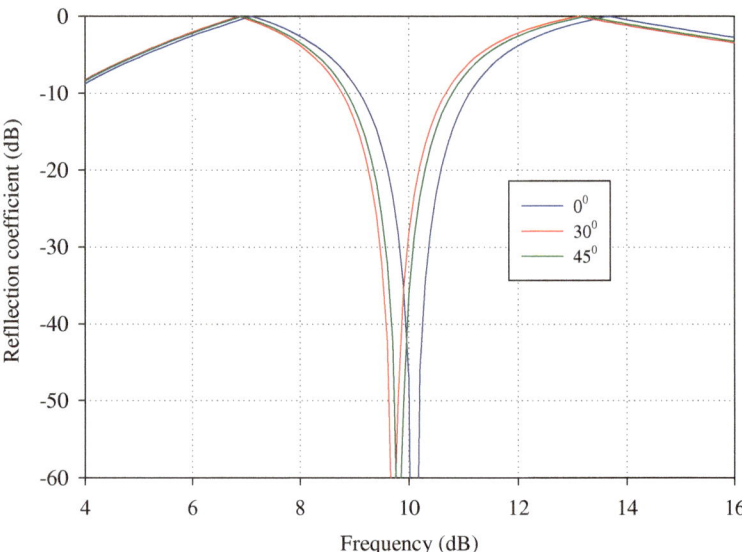

Fig. 27 Reflection characteristics of proposed DSL-FSS structure at different incidence angles (0°, 30°, and 45°) for TE polarizations

4.4 EM Performance Analysis

The input impedance and return loss of superstrate-based antenna is determined using equivalent circuit approach. For the analysis, both microstrip antenna and the FSS are designed at the frequency of 10 GHz. The input impedance of MPA loaded with superstrate is estimated using Eq. (33). The variation of real as well as imaginary parts of input impedance is studied with respect to operating frequency as shown in Fig. 28a. It is observed that the MPA loaded with FSS-based superstrate resonates at 10 GHz as it shows maximum value (48 Ω) of real parts of input impedance and minimum value (zero) of imaginary parts of input impedance at the resonance. Further, the return loss of the proposed antenna is determined using Eq. (38) and its response is studied with respect to frequency as shown in Fig. 28b. It shows minimum return loss (-11.8 dB) at the resonance frequency of 10 GHz.

The radiation characteristic of proposed antenna is estimated based on reciprocity theorem and transmission line analogy. For efficacy of this approach, the far-field radiation pattern of a rectangular MPA is validated with that of the reported results (Volakis 2007), for the antenna design parameters; height of the substrate, $h = 0.1588$ cm, ratio of height to wavelength, $h/\lambda = 0.02$, and patch aspect ratio (width to length ratio), $W/L = 1.5$. Excellent agreement is obtained between computed and reported results for both E- and H-plane pattern as shown in the Fig. 29a, b, respectively.

In order to compute the radiation characteristics of proposed antenna, the reflection characteristics of antenna is first determined based on transmission line theory for the plane wave impinging on it using Eq. (38). The radiation characteristics (E- and H-plane) of the antenna loaded with FSS-superstrate is then computed using the Eqs. (41) and (42), respectively at the frequency of 10 GHz.

The radiation characteristics of proposed antenna in E- and H-plane are compared with the conventional rectangular MPA as shown in Fig. 30a, b. It is evident that the directivity of proposed antenna is enhanced by 3.85 dB in E-plane and 4.06 dB in H-plane as compared to that of conventional MPA.

4.4.1 Performance Analysis with Air Gap Between Antenna and Superstrate

The radiation characteristics of proposed antenna is also analyzed by keeping air gap between the MPA and FSS-superstrate. This air gap is considered to be half of the operating wavelength. The directivity of the antenna is computed for E-plane as well as H-plane by using the same transmission line analogy and reciprocity theorem as discussed in the previous section. The E-plane and H-plane pattern of proposed antenna with air gap is compared with the conventional rectangular MPA as shown in Fig. 31a, b.

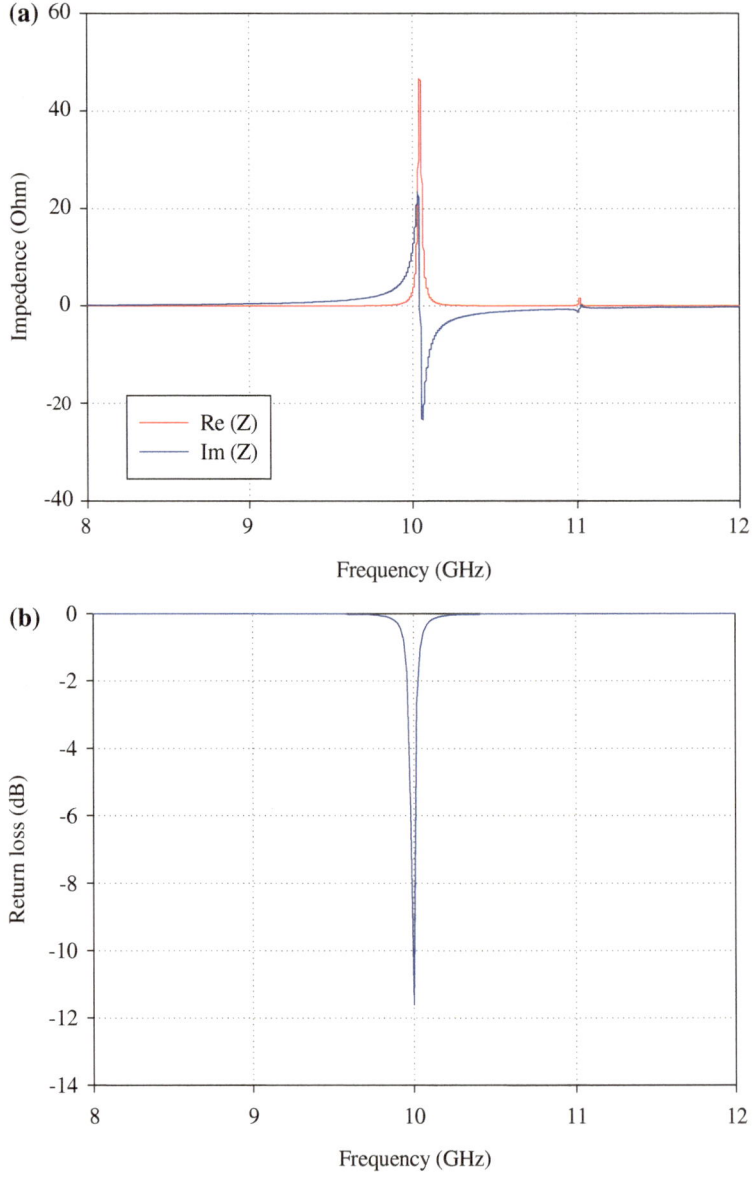

Fig. 28 EM characteristics of MPA loaded with superstrate; **a** input impedance, and **b** return loss

The directivity enhancement of antenna is observed to be 4.7 dB in *E*-plane and 4.06 in *H*-plane as compared to conventional rectangular MPA. This implies that the directivity of antenna in *E*-plane is further enhanced with air gap, while *H*-plane remained the same as that of without air gap.

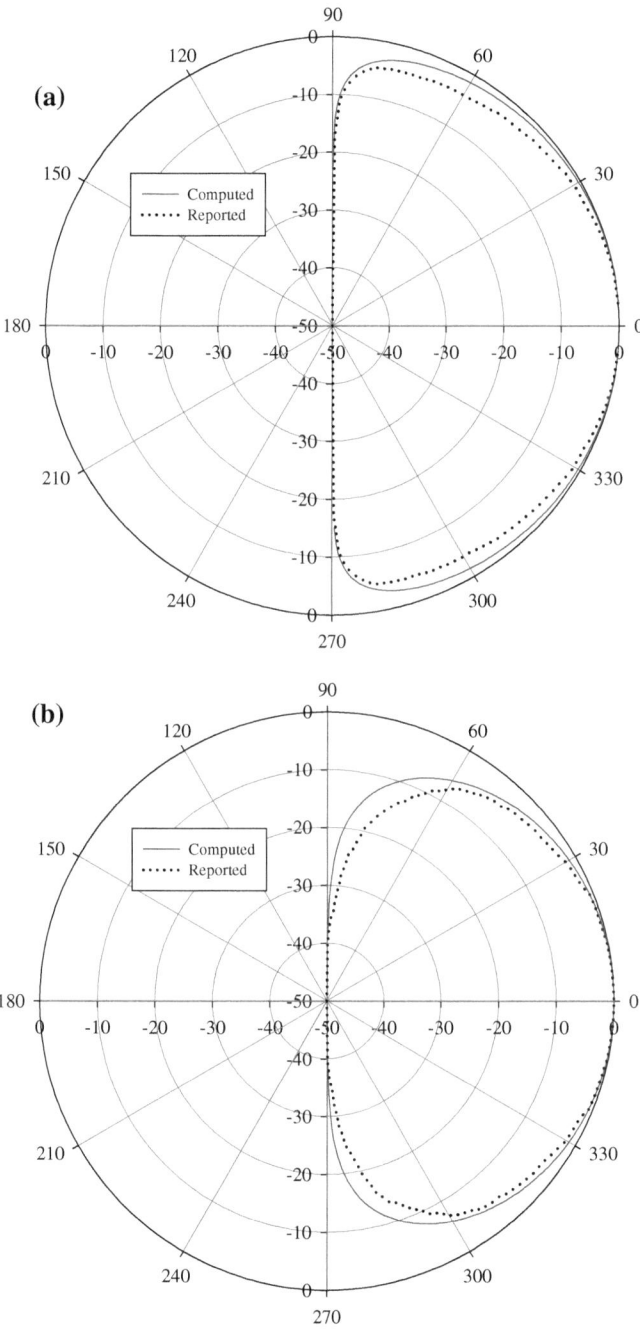

Fig. 29 Validation of **a** E-plane, and **b** H-plane pattern of rectangular MPA. *Dotted black lines* show reported results (Volakis 2007). *Solid blue lines* show computed results at CEM

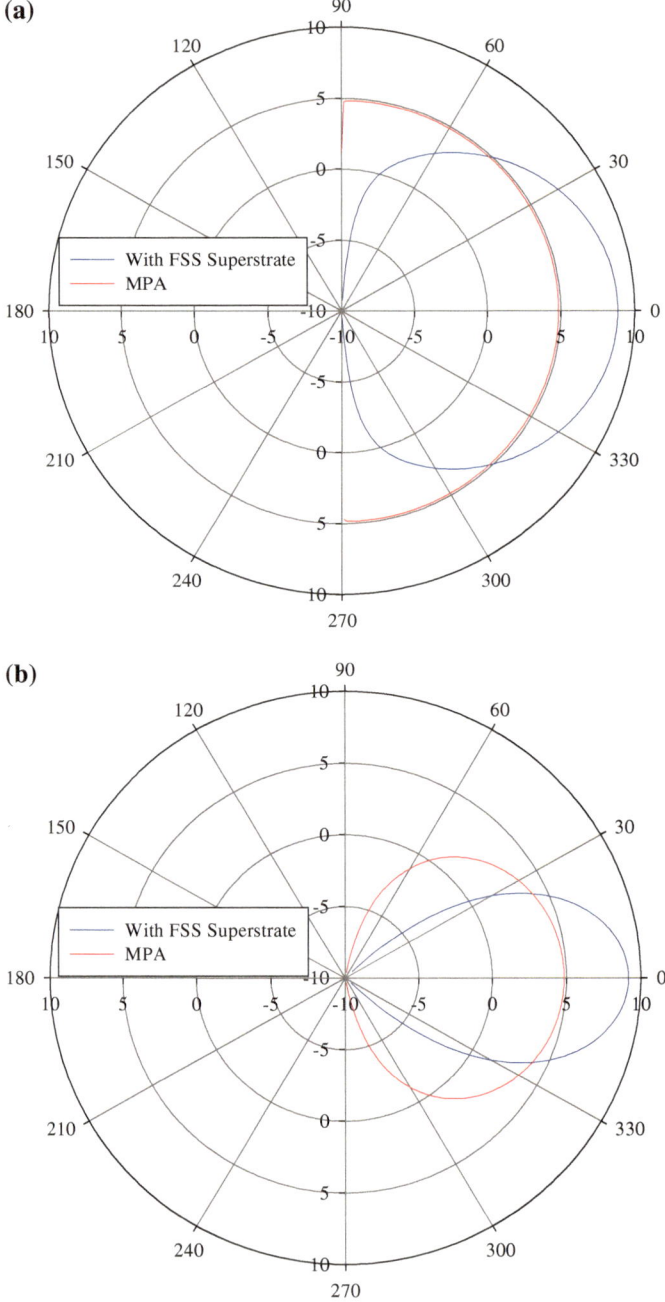

Fig. 30 Radiation pattern of rectangular MPA and MPA covered with DSL-FSS superstrate; **a** *E*-plane and **b** *H*-plane

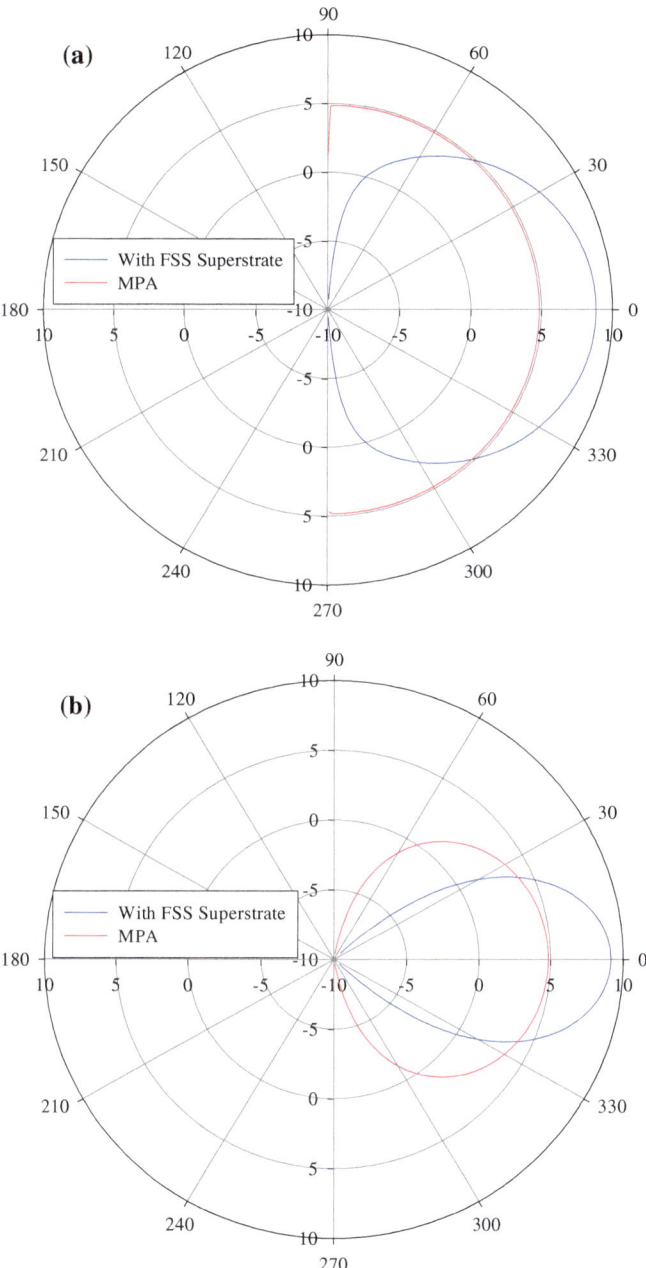

Fig. 31 Radiation pattern of rectangular microstrip antenna and MPA covered with FSS-based superstrate by keeping air gap between them; **a** *E*-plane and **b** *H*-plane

5 Summary

The present book dealt with the design and analysis of FSS-based high performance MPA. The FSS structures were used as high impedance ground plane and superstrate for the antenna. The EM analysis of a microstrip antenna loaded with FSS-based HIS has been carried out using cavity model in combination with equivalent circuit approach. The microstrip antenna was designed using cavity model, while FSS-based HIS was designed based on equivalent circuit model. For efficacy of the equivalent circuit approach, the computed results of HIS-based antenna is validated with reported results, which was obtained based on full-wave analysis method. Further, the EM performance characteristics of rectangular MPA over FSS-based HIS is studied based on equivalent circuit model. It is found that the impedance bandwidth (−10 dB) of proposed antenna is enhanced to 19.8 % as compared with that of conventional rectangular MPA (10.2 %) by loading the FSS-based HIS as ground plane. It is also revealed that the impedance bandwidth of the FSS-HIS-based antenna can be tuned by varying the geometrical parameters of FSS elements. The EM performance of proposed MPA over single square loop FSS-based HIS is also studied for different dielectric material of HIS substrate. It is found that the bandwidth of HIS-based antenna can be tuned by changing the dielectric material of HIS substrate.

Further, the EM analysis of a microstrip antenna loaded with FSS superstrate has been carried out in this book by using transmission line equivalent circuit model and reciprocity theorem. For efficacy of the approach, the far-field pattern (E- and H-plane) of a rectangular MPA is validated with that of reported results. Excellent agreement is obtained between computed and reported results. The MPA covered with DSL-FSS superstrate exhibited directivity enhancement of 3.85 dB in E-plane and 4.06 dB in H-plane as compared to that of conventional MPA. Further, the directivity of proposed antenna has been enhanced by keeping air-gap between antenna and FSS layer, which is observed to be 4.7 dB in E-plane and 4.06 in H-plane as compared to the conventional rectangular MPA. Since superstrate has been designed using DSL-FSS which will reject the impinging signal on antenna structure outside the operating band, and hence it will reduce out-of-band RCS of proposed antenna. However, estimation of out-of-band RCS has not been discussed in this context. Thus, the proposed FSS-superstrate-based antenna may find potential applications at low observable platforms. To conclude, the topics discussed in this book would serve as the building block for designing the FSS-based low observable antennas.

References

Alexopoulos, N. G. and D. R. Jackson. 1984. Fundamental superstrate (cover) effects on printed circuit antennas. *IEEE Transaction on Antennas and Propagation* AP-32: 807–816.
Bahl, I.J., and P. Bhartia. 1980. *Microstrip antennas*. Dedham: Artech House.
Balanis, C.A. 1997. *Antenna theory: Analysis and design*. New York: Wiley.

Carver, K.R., and J.S. Mink. 1981. Microstrip antenna technology. *IEEE Transaction on Antennas and Propagation* 29(1): 2–24.

Chen, J. C. and P.H. Stanton. 1991. Theoretical and experimental results for a thick skew-grid FSS with rectangular apertures at oblique incidence. *Proceedings of IEEE Antennas and Propagation Society International Symposium* 3: 1866–1869.

Chiu, S. C. and S. Y. Chen. 2011. High-gain circularly polarized resonant cavity antenna using FSS superstrate. *IEEE International Symposium on Antennas and Propagation* (APSURSI), Spoken, WA, pp. 2242–2245.

Costa, F., and A. Monarchio. 2012a. Closed-form analysis of reflection losses in microstrip reflectarray antennas. *IEEE Transaction on Antennas and Propagation* 60(10): 4650–4660.

Costa, F., and A. Monarchio. 2012b. A frequency selective radome with wideband absorbing properties. *IEEE Transactions on Antennas and Propagation* 60: 2740–2747.

Costa, F., A. Monarchio, and G. Manara. 2012. Efficient analysis of frequency-selective surface by simple equivalent-circuit model. *IEEE Antennas and Propagation Magazine* 54(4): 35–48.

Derneryed, A.G., and A.G. Lind. 1979. Extended analysis of rectangular microstrip resonator antennas. *IEEE Trans. Antennas Propag.* 27: 846–849.

Foroozesh, A., and L. Shafai. 2006. 2-D truncated periodic leaky-wave antennas with reactive impedance surface ground. *Proceedings of IEEE AP-S International Symposium*, Albuquerque, NM, pp. 15–18

Foroozesh, A., and L. Shafai. 2010. Investigation into the effects of the patch-type FSS superstrate on the high-gain cavity resonance antenna design. *IEEE Transactions on Antennas and Propagation* 58(2): 258–270.

Genovesi, S., F. Costa, and A. Monorchio. 2012. Low-profile array with reduced radar cross section by using hybrid frequency selective surfaces. *IEEE Transaction on Antennas and Propagation* 60(5): 2327–2335.

Gupta, G., and K. Gururaj. 2013. Neurostimulation device having frequency selective surface to prevent electromagnetic interference during MRI. *Patent US20130274829 A1.*

Gustafsson, M., A. Karlsson, A.P.P. Rebelo, and B. Widenberg. February 2005. Design of frequency selective windows for improved indoor outdoor communication. *Department of Electroscience, Lund Institute of Technology*, Sweden, Electromagnetic Theory Tech. Rep. TEAT-7132, 16 p.

Hosseini, M., and M. Hakkak. 2008. Characteristics estimation for Jerusalem cross–based artificial magnetic conductors. *IEEE Antennas and Wireless Propagation Letters* 7: 58–61.

Hosseinipanah, M., and Q. Wu. 1985. Equivalent circuit model for designing of Jerusalem cross based artificial magnetic conductors. *Radio Engineering* 18(4).

Jackson, D.R., and N.G. Alexopoulos. 1985. Gain enhancement methods for printed circuit antennas. *IEEE Transactions on Antennas and Propagation* AP-33(9): 976–987.

Kanaujia, B.K., and B.R. Viswakarma. 2006. Reactively loaded annular microstrip antenna for multiband operation. *Indian Journal of Radio and Space Physics* 35: 122–128.

Lee, C.K., and R.J. Langley. 1985. Equivalent-circuit models for frequency selective surfaces at oblique angles of incidence. *IEEE Proceedings of Microwaves, Antennas and Propagation* H-132(6): 395–399.

Lee, Y.J., J. Yeo, R. Mittra, and W.S. Park. 2004. Design of a high-directivity electromagnetic band gap (EBG) resonator antenna using a frequency-selective surface (FSS) superstrate. *Microwave and Optical Technology Letters* 43(6): 462–467.

Li, H., B.-Z. Wang, G. Zheng, and W. Shao. 2010. A reflectarray antenna backed on FSS for low RCS and high radiation performances. *Progress In Electromagnetics Research C* 15: 145–155.

Loui, H, 2006. Modal analysis and design of compound gratings and frequency selective surfaces. Ph.D. Thesis, *Department of Electrical and Computer Engineering, University of Colorado*, Boulder, 139 p.

Lu, B., X. Gong, J. Ling, and H.W. Yuan. 2009. Radar cross section reduction of antennas using stop-band frequency selective surfaces. *Microwave Journal* 16 p.

Luo, X.F., P.T. Teo, A. Qing, and C.K. Lee. 2005. Design double-square-loop frequency-selective surfaces using differential evolution strategy coupled with equivalent-circuit model. *Microwave and Optical Technology Letters* 44(2): 159–162.

Monavar, F.M., and M. Komjani. 2011. Bandwidth enhancement of microstrip patch antenna using Jerusalem cross-shaped frequency selective surfaces by invasive weed optimization approach. *Progress in Electromagnetics Research* 121: 103–120.

Narayan, S., K. Prasad, R.U. Nair, and R.M. Jha. 2012. A novel EM analysis of double-layered thick FSS based on MM-GSM technique for radome applications. *Progress in Electromagnetics Research Letters* 28: 53–62.

Philippakis, M., C. Martel, D. Kemp, R. Allan, M. Clift, S. Massey, S. Appleton, W. Damerell, C. Burton, and E.A. Parker. March 2004. Application of FSS structures to selectively control the propagation of signals into and out of buildings. Technical report: 2004-0072, ERA Technology Ltd., Cleeve road, Leatherhead, U.K.

Pirhadi, A., M. Hakkak, and F. Keshmiri. 2006. Bandwidth enhancement of the probe fed microstrip antenna using frequency selective surface as electromagnetic bandgap superstrate. *Progress in Electromagnetics Research* 61: 215–230.

Pirhadi, A., F. Keshmiri, M. Hakkak, and M. Tayarani. 2007. Analysis and design of dual band high directivity EBG resonator antenna using square loop FSS as superstrate layer. *Progress in Electromagnetic Research* PIER 70: 1–20.

Pirhadi, A., H. Bahrami, and J. Nasri. 2012. Wideband high directivity aperture coupled microstrip antenna design by using a FSS superstrate layer. *IEEE Transactions on Antennas and Propagation* 60(4): 2101–2106.

Rodes, E., M. Diblanc, E. Arnaud, T. Monédière, and B. Jecko. 2007. Dualband EBG resonator antenna using a single-layer FSS. *IEEE Antennas Wireless Propagation Letter* 6: 368–371.

Volakis, J.L. 2007. *Antenna engineering handbook*, 4th ed. New York: McGraw-Hill.

Yang, H.-H., X.-Y. Cao, Q.-R. Zheng, J.-J. Ma, and W.-Q. Li. 2013. Broadband RCS reduction of microstrip patch antenna using bandstop frequency selective surface. *Radio Engineering* 22(4): 1275–1280.

Yeo, J., R. Mittra, and S. Chahavany. 2002. A GA-based design of electromagnetic bandgap structures utilizing frequency selective surfaces for bandwidth enhancement of microstrip antennas. *IEEE International Symposium on Antennas and Propagation* 2: 400–403. (San Antonio, TX).

About the Book

This book focuses on performance enhancement of printed antennas using frequency selective surfaces (FSS) technology. The growing demand of stealth technology in strategic areas requires high-performance low-RCS (radar cross section) antennas. Such requirements may be accomplished by incorporating FSS into the antenna structure either in its ground plane or as the superstrate, due to the filter characteristics of FSS structure. In view of this, a novel approach based on FSS technology is presented in this book to enhance the performance of printed antennas including out-of-band structural RCS reduction. In this endeavor, the EM design of microstrip patch antennas (MPA) loaded with FSS-based (i) high impedance surface (HIS) ground plane, and (ii) the superstrates are discussed in detail. The EM analysis of proposed FSS-based antenna structures have been carried out using transmission line analogy, in combination with the reciprocity theorem. Further, various types of novel FSS structures are considered in designing the HIS ground plane and superstrate for enhancing the MPA bandwidth and directivity. The EM design and performance analyses of FSS-based antennas are explained here with the appropriate expressions and illustrations.

© The Author(s) 2016   41
S. Narayan et al., *Frequency Selective Surfaces based High Performance*
Microstrip Antenna, SpringerBriefs in Computational Electromagnetics,
DOI 10.1007/978-981-287-775-8

Author Index

A

Alexopoulos, N.G., 23–24, 29
Allan, R., 1
Appleton, S., 1
Arnaud, E., 24

B

Bahl, I.J., 8
Bahrami, H., 1, 2, 23–24
Balanis, C.A., 5
Bhartia, P., 8
Burton, C., 1

C

Cao, X.-Y., 2
Carver, K.R., 5
Chahavany, S., 2
Chen, J.C., 3
Chen, S.Y., 2
Chiu, S.C., 2
Clift, M., 1
Costa, F., 4, 9–10, 26

D

Damerell, W., 1
Derneryed, A.G., 8
Diblanc, M., 24

F

Foroozesh, A., 2, 23

G

Genovesi, S., 2, 4
Gong, X., 2
Gupta, G., 1
Gururaj, K., 1
Gustafsson, M., 3

H

Hakkak, M., 2, 5, 8, 23
Hosseini, M., 5, 8
Hosseinipanah, M., 5, 8, 12, 14

J

Jackson, D.R., 23–24, 28, 29
Jecko, B., 24

K

Kanaujia, B.K., 7
Karlsson, A., 3
Kemp, D., 1
Keshmiri, F., 2, 23
Komjani, M., 2, 5, 14

L

Langley, R.J., 26–27
Lee, C.K., 24, 26–27
Lee, Y.J., 2
Li, H., 1, 2
Li, W.-Q., 2
Lind, A.G., 8
Ling, J., 2
Loui, H., 3
Lu, B., 2, 4
Luo, X.F., 24, 25, 31

M

Ma, J.-J., 2
Manara, G., 4, 9–10
Martel, C., 1
Massey, S., 1
Mink, J.S., 5
Mittra, R., 2
Monarchio, A., 4, 9–10
Monavar, F.M., 2, 5, 14
Monédière, T., 24

© The Author(s) 2016
S. Narayan et al., *Frequency Selective Surfaces based High Performance Microstrip Antenna*, SpringerBriefs in Computational Electromagnetics, DOI 10.1007/978-981-287-775-8

N
Nair, R.U., 4
Nasri, J., 1, 24

P
Park, W.S., 2
Parker, E.A., 1
Philippakis, M., 1
Pirhadi, A., 1, 2, 23–24
Prasad, K., 4

Q
Qing, A., 24

R
Rebelo, A.P.P., 3
Rodes, E., 24

S
Shafai, L., 2, 24
Shao, W., 1
Stanton, P.H., 3

T
Tayarani, M., 2
Teo, P.T., 24

V
Viswakarma, B.R., 7
Volakis, J.L., 12, 16, 29, 33

W
Wang, B.-Z., 1
Widenberg, B., 3
Wu, Q., 5, 8, 12–14

Y
Yang, H.-H., 2
Yeo, J., 2
Yuan, H.W., 2

Z
Zheng, G., 1–2
Zheng, Q.-R., 2

Subject Index

A
Antennas, 1, 2, 4
Artificial magnetic conductor, 2, 3, 5

B
Bandwidth, 1–2, 4–5, 15, 16, 20, 23, 38
Beamwidth, 16, 20, 23

C
Capacitive FSS, 3
Cavity model, 5, 7, 11, 24, 30, 38
Complex relative permittivity, 9

D
Directivity, 2, 23–24, 31, 33–38

E
Effective length, 6
Electromagnetic interference (EMI), 1
Equivalent circuit model, 2, 5, 10, 11–13, 20, 38

F
Frequency selective surfaces (FSS), 1, 23
Fringing phenomenon, 5
Full-wave analysis method, 2, 5, 38

G
Ground plane, 1–5, 8, 11, 12, 15–19, 21, 28, 38

H
High impedance ground plane, 1–2, 4, 8, 38
High impedance surface (HIS), 9–11

I
Inductive FSS, 3

J
Jerusalem crossed FSS (JC-FSS), 2, 5, 8–9, 11, 16

M
Microstrip patch antenna, 2, 5, 11, 19, 24, 25, 28, 38

P
Perfect magnetic conductor (PMC), 5
Planar antennas, 1, 4
Propagation constant, 28–29

R
Radar cross section (RCS), 1, 2, 4, 23, 38
Radiation characteristics, 2, 3, 23, 33
Radiation pattern, 18, 28, 29, 33
Reciprocity theorem, 16, 20, 24, 28, 29, 33
Rectangular MPA, 7, 8, 11, 15, 16, 20, 24, 30, 38
Reflection coefficient, 8, 10, 16, 29–30
Refractive index, 29–30
Return loss, 10, 11, 15, 21, 22, 28, 30

S
Single square loop, 5, 18, 20, 38
Superstrate, 1–2, 23–25, 27–31, 33–34

T
Thick filter, 3
Thin filter, 3
Transmission efficiency, 3, 31
Transmission line analogy, 16, 24, 29, 33

© The Author(s) 2016

S. Narayan et al., *Frequency Selective Surfaces based High Performance Microstrip Antenna*, SpringerBriefs in Computational Electromagnetics, DOI 10.1007/978-981-287-775-8